Selected Papers from the 1st International Online Conference on Nanomaterials

Selected Papers from the 1st International Online Conference on Nanomaterials

Special Issue Editors

Ana María Díez-Pascual
Guanying Chen

MDPI • Basel • Beijing • Wuhan • Barcelona • Belgrade • Manchester • Tokyo • Cluj • Tianjin

Special Issue Editors

Ana María Díez-Pascual   Guanying Chen
Alcalá University   Harbin Institute of Technology
Spain   China

Editorial Office
MDPI
St. Alban-Anlage 66
4052 Basel, Switzerland

This is a reprint of articles from the Special Issue published online in the open access journal *Nanomaterials* (ISSN 2079-4991) (available at: https://www.mdpi.com/journal/nanomaterials/special_issues/iocn_1).

For citation purposes, cite each article independently as indicated on the article page online and as indicated below:

LastName, A.A.; LastName, B.B.; LastName, C.C. Article Title. *Journal Name* **Year**, *Article Number*, Page Range.

ISBN 978-3-03928-534-1 (Hbk)
ISBN 978-3-03928-535-8 (PDF)

Contents

About the Special Issue Editors

Ana María Díez-Pascual graduated in Chemistry in 2001 (awarded Extraordinary Prize) at the Complutense University (Madrid, Spain), where she completed her Ph.D. (2002–2005) on dynamic and equilibrium properties of fluid interfaces under the supervision of Prof. Rubio. In 2005, she worked at the Max Planck Institute of Colloids and Interfaces (Germany) with Prof. Miller on the rheological characterization of water-soluble polymers. During 2006–2008, she was a postdoctoral researcher at the Physical Chemistry Institute of the RWTH-Aachen University (Germany), where she worked on the layer-by-layer assembly of polyelectrolyte multilayers onto thermoresponsive microgels. Then, she moved to the Institute of Polymer Science and Technology (Madrid, Spain) and participated in a Canada–Spain joint project to develop carbon nanotube (CNT)-reinforced epoxy and polyetheretherketone composites for transport applications. Currently, she is a postdoctoral researcher at Alcalá University (Madrid, Spain) focused on the development of polymer/nanofiller systems for biomedical applications. She has published over 100 SCI articles (97% in Q1 journals). She has an H-index of 41 and more than 3500 total citations. She has published 21 book chapters, 2 monographies, and edited 3 books, and is the first author of an international patent. She has been invited to impart seminars at prestigious international research centers (i.e., Max Planck in Germany, NRC in Canada, School of Materials in Manchester, U.K.). She was awarded the TR35 2012 prize by the Massachusetts Technological Institute (MIT) for her innovative work in the field of nanotechnology.

Guanying Chen received his Ph.D. degree in Optics in 2009 from Harbin Institute of Technology (HIT), China. After that, he became an assistant professor (2009), associate (2012), and full professor (2013) at School of Chemistry and Chemical Engineering at HIT. He also holds a joint position as senior research scientist (since 2012) at the Institute for Lasers, Photonics and Biophotonics (ILPB), State University of New York in Buffalo. His interests include lanthanide luminescence, optical bioimaging, biosensing, and solar cells. He has published more than 100 peer-reviewed papers in prestigious international journals, 8 book chapters, and 4 issued patents, and delivered 60 invited talks in internal conferences and universities. He also serves as an Editorial Board member for the journals Scientific Reports, Nanomaterials, Crystals, and Chinese Journal of Luminescence, and as the Guest Editor for Theranostics.

 *nanomaterials*

Editorial

Selected Papers from the 1st International Online Conference on Nanomaterials

Ana María Díez-Pascual [1],* and Guanying Chen [2],*

[1] Department of Analytical Chemistry, Physical Chemistry and Chemical Engineering, Faculty of Sciences, University of Alcalá, 28805 Alcalá de Henares, Madrid, Spain
[2] School of Chemistry and Chemical Engineering, Harbin Institute of Technology, Harbin 150001, China
* Correspondence: am.diez@uah.es (A.M.D.-P.); chenguanying@hit.edu.cn (G.C.); Tel.: +34-918-856-430 (A.M.D.-P.); +86-451-86403809 (G.C.)

Received: 1 July 2019; Accepted: 12 July 2019; Published: 17 July 2019

After decades of intense research, nanomaterials are now an integral part of many applications and enjoy the attention of a large research community. Intrinsically multidisciplinary, research activities are spanning from engineering, over physics and chemistry, to biology and medicine. Nanomaterials, as the name indicates, are extremely tiny, less than a millionth of a meter in size. They encompass exceptional physical and chemical features which provide them enhanced properties compared to their macroscopic counterparts, such as higher reactivity and strength, superior thermal and electrical characteristics as well as functionality. These advantages have led to nanomaterials being included in a broad range of consumer products. The transport, electronics, cosmetics, healthcare, and sport industries all benefit from nanotechnology advances. Novel fields have also appeared, such as nanomedicine, which plans to drastically change our future ability to treat disease.

This Special Issue compiles five selected papers from the Proceedings of the 1st International Online Conference on Nanomaterials, held 1–15 September 2018 on sciforum.net, an online platform for hosting scholarly e-conferences and discussion groups. It targets a broad readership of physicists, chemists, materials scientists, biologists, environmentalists, and nanotechnologists, and provides interesting examples of the most recent advances in the synthesis, characterization, and applications of nanomaterials. The papers present very different types of nanomaterials, such as double hydrophilic branched copolymers of poly(N,N-dimethyl acrylamide) and poly(ethylene oxide) [1], carbon nano-dots (CNDs)/poly(vinyl alcohol) (PVA) nanocomposites [2], polyacrylamide/SiO_2 hydrogel nanocomposites [3], silicon quantum dots (Si QDs) and iron oxide (α-Fe_2O_3) nanoparticles [4], and hexamethylene diisocyanate (HDI)-functionalized graphene oxide (GO) [5]. In the paragraphs that follow, a concise overview of each of the published articles is provided in order to attract the interest of potential readers.

Recently, hydrogels have emerged as ideal candidates for numerous biomedical and biotechnological applications due to their unique physical and biochemical properties [6]. They comprise very porous and hydrated networks that enable cell encapsulation for tissue engineering, the loading and release of bioactive molecules for drug delivery, and wound dressing and biosensing applications [7]. Nonetheless, their poor mechanical performance limits their use in certain applications, and strong effort has been carried out on improving their properties via incorporation of nanofillers such as inorganic nanoparticles [8], though the precise mechanisms behind such enhancements are not fully understood yet. To get more insight into the role of nanoparticles on the mechanical properties of hydrogel nanocomposites, Zaragoza et al. [3] synthesized chemically crosslinked polyacrylamide hydrogels reinforced with SiO_2 nanoparticles as a model system of study. Rheological experiments revealed that the improvements induced by means of the nanoparticles surpass the maximum modulus that can be attained via simply chemical crosslinking. Furthermore, results demonstrate that the concentrations of the nanoparticle, monomer, and chemical crosslinker play a key role in mechanical improvement, which is crucial for their use in a wide range of applications.

Carbon nano-dots (CNDs) represent a novel class of carbon-based nanomaterials with quasi-spherical shape and ultra-small size of ~10 nm, that has attracted a lot of attention among studies, owed to their valuable characteristics such as cheapness, biodegradability, strong and broad optical absorption, and high chemical stability [9]. These extraordinary properties make them useful for a variety of fields including biosensing, bioimaging, drug delivery, optoelectronics, photovoltaics, and photocatalysis. In particular, the rich optical and electronic properties of CNDs including efficient light harvesting, tunable photoluminescence, and superior photoinduced electron transfer have involved considerable interest in different photocatalytic applications. The addition of CNDs to polymeric matrices is currently under intense study since they display great potential for light emitting diodes (LEDs), flexible electronic displays, and other optoelectronic applications [10]. In this context, Aziz et al. [2] investigated the effect of CNDs on the UV absorption spectra of PVA-based nanocomposites prepared via solution casting. The existence of CNDs of different sizes well dispersed throughout the matrix was proved via microscopic observations. In addition, the gradual increase of the refractive index with increasing CND concentration corroborated the homogeneous distribution of the carbon nanofillers all over the host PVA. The Infrared spectra and X-ray diffraction data demonstrated the complex formation between PVA and CNDs. Further, the CNDs caused strong absorptions at 280 and 330 nm assigned to n-π^* and π-π^* transitions, as well as a reduction in the optical bandgap.

Another highly interesting carbon-based nanomaterial is graphene oxide (GO), the oxidized form of graphene, which exhibits exceptional properties including high mechanical strength, optical transparency, amphiphilicity, and surface functionalization capability [11]. Even though, its insolubility in non-polar and polar aprotic solvents limits certain applications. To work out this matter, new functionalization approaches are required [12]. In this regard, Luceño-Sanchez et al. [5] prepared and characterized a series of hexamethylene diisocyanate (HDI)-functionalized GO. Several reaction conditions were tested to maximize the degree of modification, and comprehensive characterizations were carried out by means of elemental analysis, Infrared, and Raman spectroscopies to verify the accomplishment of the functionalization reaction. The surface morphology of the modified samples was explored by microscopic techniques, which showed a rise in the sheet thickness with increasing the level of modification. The HDI-GO was found to be more hydrophobic in nature than neat GO and was easily suspended in polar aprotic solvents such as N,N-dimethylformamide (DMF), N-methylpyrrolidone (NMP) and dimethyl sulfoxide (DMSO), as well as in low polar/non-polar solvents like tetrahydrofuran (THF), chloroform and toluene. Further, the dispersibility increased with increasing degree of modification. Besides, it was found that the covalent bonding of HDI enhances the thermal stability of GO due to chemical crosslinking between neighboring sheets, which is advantageous for long-term electronics and electrothermal device applications. These HDI-GO samples are perfect candidates as nanofillers for the development of high-performance GO-based polymeric nanocomposites [13,14].

The use of nanomaterials as optical sensors is emerging as a new frontier in nanoscience, which enables to detect hint analyte of varying kinds (temperature, oxygen molecules, peroxide, disease biomarkers, etc.) at high precision [15]. Moreover, nanomaterials optical sensing can empower remote and noninvasive evaluation of microenvironment at a nanometric resolution, providing unprecedented opportunities for quantifying analytes of interest, such as humidity [16]. Humidity is present everywhere and is one of the most important physical parameters in semiconductors, electronics, food processing, and pharmaceutical industries wherever the quality of products is influenced by water molecules. In this context, Lazarova et al. described a thin film humidity optical sensor with nanometric thicknesses, 100–400 nm, prepared from double hydrophilic copolymers of complex branched structures (containing poly(N,N-dimethyl acrylamide) and poly(ethylene oxide) blocks) [1]. The polymer thin films were deposited on top-covered Bragg stacks with a sputtered Au–Pd film (30 nm) that bring color for the colorless polymer/glass system, thus enabling transmittance measurements for humidity sensing. The humidity content was quantitatively studied by calculating the color coordinate alternation using the measured spectra of transmittance. This work provides a paradigm of using top covered Bragg stacks and polymer/metal thin film structures as sensitive humidity optical sensors.

Nanomedicine represents the next era for personalized disease prevention and treatment with better performance and fewer side effects. One important direction of nanomedicine involves the use of nanocarriers to deliver drugs or prodrugs to intended sites in a stimuli-controlled manner for precise theranostic treatment and actuation in a living organism [17]. Among various delivery ways, oral drug administration is still the preferred route in this regard, with advantages including patient comfort, reduced chances of infection, and minimal invasiveness. However, orally administered drugs usually have to be assimilated via the gastrointestinal (GI) tract, which then crosses the small intestinal barrier to reach vascular circulation for intended sites [18]. As such, an in vitro model to assess intestinal absorption is of paramount importance to mimic the conditions of drug-loaded nanocarriers in GI. To do so, Strugari et al. co-cultured Caco-2 (cat. no. CRL-2102) cells and human adenocarcinoma line HT-29 preconditioned in methotrexate (MTX) on a Transwell® systems in a monolayer [4]. This co-culture model allows for modulating intercellular junction geometry, thereby fine-tuning the effective permeability of the monolayer by simply adjusting the initial cell seeding ratio of Caco-2/HT29-MTX (7:3 and 5:5 in the work). The monolayer integrity is assessed by measuring transepithelial electrical resistance (TEER) using chopstick electrodes inserted in the apical (AP) and basolateral (BL) compartments of the well. This model was exposed to non-cytotoxic concentration levels (20 µg/mL) of Si QDs and α-Fe$_2$O$_3$ nanoparticles, quantitatively evaluating their ability to penetrate through the intestinal mucus barrier, and their influence on the morphological alterations of the Caco-2/HT29-MTX. However, the obtained results showed that the current surface of these nanoparticle suspensions prevented them from diffusing across the intestinal model, demonstrating the importance of nanoparticle surface functionalization to enable them to cross the intestinal barriers.

These papers published in the special issue of the 1st International Online Conference on Nanomaterials (IOCON) showcase the important roles of nanomaterials to influence science and technology and human healthcare. We hope that more and more scientists can join the open-access ICON forum in the future to facilitate the advancement of nanoscience together.

Acknowledgments: Ana Díez-Pascual wishes to knowledge the Ministerio de Economía y Competitividad (MINECO) for a "Ramón y Cajal" Research Fellowship cofinanced by the EU. Guanying Chen acknowledges the funding support from National Natural Science Foundation of China (51672061) and the Fundamental Research Funds for the Central Universities, China (HIT.BRETIV.201503) for his researches on nanomaterials.

Conflicts of Interest: The authors declare no conflicts of interest.

References

1. Lazarova, K.; Christova, D.; Georgiev, R.; Georgieva, B.; Babeva, T. Optical Sensing of Humidity Using Polymer Top-Covered Bragg Stacks and Polymer/Metal Thin Film Structures. *Nanomaterials* **2019**, *9*, 875. [CrossRef] [PubMed]
2. Aziz, S.B.; Hassan, A.Q.; Mohammed, S.J.; Karim, W.O.; Kadir, M.F.Z.; Tajuddin, H.; Chan, N.N.M.Y. Structural and Optical Characteristics of PVA:C-Dot Composites: Tuning the Absorption of Ultra Violet (UV) Region. *Nanomaterials* **2019**, *9*, 216. [CrossRef] [PubMed]
3. Zaragoza, J.; Fukuoka, S.; Kraus, M.; Thomin, J.; Asuri, P. Exploring the Role of Nanoparticles in Enhancing Mechanical Properties of Hydrogel Nanocomposites. *Nanomaterials* **2018**, *8*, 882. [CrossRef] [PubMed]
4. Strugari, A.F.; Stan, M.S.; Gharbia, S.; Hermenean, A.; Dinischiotu, A. Characterization of Nanoparticle Intestinal Transport Using an In Vitro Co-Culture Model. *Nanomaterials* **2019**, *9*, 5. [CrossRef] [PubMed]
5. Luceño-Sánchez, J.A.; Maties, G.; Gonzalez-Arellano, C.; Diez-Pascual, A.M. Synthesis and Characterization of Graphene Oxide Derivatives via Functionalization Reaction with Hexamethylene Diisocyanate. *Nanomaterials* **2018**, *8*, 870. [CrossRef] [PubMed]

6. Liaw, C.Y.; Ji, S.; Guvendiren, M. Engineering 3D hydrogels for personalized in vitro human tissue models. *Adv. Healthc. Mater.* **2018**, *7*, 1701165. [CrossRef] [PubMed]

7. Peppas, N.A.; Huang, Y.; Torres-Lugo, M.; Ward, J.H.; Zhang, J. Physicochemical foundations and structural design of hydrogels in medicine and biology. *Annu. Rev. Biomed. Eng.* **2000**, *2*, 9–29. [CrossRef] [PubMed]

8. Tjong, S.C. Structural and mechanical properties of polymer nanocomposites. *Mater. Sci. Eng. R Rep.* **2006**, *53*, 73–197. [CrossRef]

9. Wang, R.; Lu, K.-Q.; Tang, Z.-R.; Xu, Y.-J. Recent progress in carbon quantum dots: Synthesis, properties and applications in photocatalysis. *J. Mater. Chem. A* **2017**, *5*, 3717–3734. [CrossRef]

10. Kovalchuk, A.; Huang, K.; Xiang, C.; Martí, A.A.; Tour, J.M. Luminescent Polymer Composite Films Containing Coal-Derived Graphene Quantum Dots. *ACS Appl. Mater. Interfaces* **2015**, *7*, 26063–26068. [CrossRef] [PubMed]

11. Dreyer, D.R.; Park, S.; Bielawski, C.W.; Ruoff, R.S. The chemistry of graphene oxide. *Chem. Soc. Rev.* **2010**, *39*, 228–240. [CrossRef] [PubMed]

12. Díez-Pascual, A.M.; Luceño Sánchez, J.A.; Peña Capilla, R.; García Díaz, P. Recent Developments in Graphene/Polymer Nanocomposites for Application in Polymer Solar Cells. *Polymers* **2018**, *10*, 217. [CrossRef] [PubMed]

13. Luceño Sánchez, J.A.; Peña Capilla, R.; Díez-Pascual, A.M. High-Performance PEDOT:PSS/Hexamethylene Diisocyanate-Functionalized Graphene Oxide Nanocomposites: Preparation and Properties. *Polymers* **2018**, *10*, 1169. [CrossRef] [PubMed]

14. Luceño Sánchez, J.A.; Díez-Pascual, A.M.; Peña Capilla, R.; García Díaz, P. The Effect of Hexamethylene Diisocyanate-Modified Graphene Oxide as a Nanofiller Material on the Properties of Conductive Polyaniline. *Polymers* **2019**, *11*, 1032. [CrossRef] [PubMed]

15. Hao, S.; Chen, G.; Yang, C. Sensing using upconversion nanoparticles. *Theranostics* **2013**, *3*, 331–345. [CrossRef] [PubMed]

16. Shi, J.; Zhu, Y.; Zhang, X.; Baeyens, W.; Campana, A.M.G. Recent developments in nanomaterial optical sensors. *TrAC Trends Anal. Chem.* **2004**, *23*, 351–360. [CrossRef]

17. Chen, G.; Roy, I.; Yang, C.; Prasad, P. Nanochemistry and nanomedicine for nanoparticle-based diagnostics and therapy. *Chem. Rev.* **2016**, *116*, 2826–2885. [CrossRef] [PubMed]

18. Zhang, N.; Ping, Q.; Huang, G.; Xu, W.; Cheng, Y.; Hang, X. Lectin-modified solid lipid nanoparticles as carriers for oral administration of insulin. *Int. J. Pharm.* **2006**, *327*, 153–159. [CrossRef] [PubMed]

Article

Optical Sensing of Humidity Using Polymer Top-Covered Bragg Stacks and Polymer/Metal Thin Film Structures

Katerina Lazarova [1,*], **Darinka Christova** [2], **Rosen Georgiev** [1], **Biliana Georgieva** [1] and **Tsvetanka Babeva** [1,*]

[1] Institute of Optical Materials and Technologies "Acad. J. Malinowski", Bulgarian Academy of Sciences, Akad. G. Bonchev str., bl. 109, 1113 Sofia, Bulgaria; rgeorgiev@iomt.bas.bg (R.G.); biliana@iomt.bas.bg (B.G.)

[2] Institute of Polymers, Bulgarian Academy of Sciences, Akad. G. Bonchev Str., bl. 103-A, 1113 Sofia, Bulgaria; dchristo@polymer.bas.bg

* Correspondence: klazarova@iomt.bas.bg (K.L.); babeva@iomt.bas.bg (T.B.); Tel.: +359-02-979-3526 (K.L.); +359-02-872-0073 (T.B.)

Received: 7 May 2019; Accepted: 7 June 2019; Published: 10 June 2019

Abstract: Thin films with nanometer thicknesses in the range 100–400 nm are prepared from double hydrophilic copolymers of complex branched structures containing poly(*N*,*N*-dimethyl acrylamide) and poly(ethylene oxide) blocks and are used as humidity sensitive media. Instead of using glass or opaque wafer for substrates, polymer thin films are deposited on Bragg stacks and thin (30 nm) sputtered Au–Pd films thus bringing color for the colorless polymer/glass system and enabling transmittance measurements for humidity sensing. All samples are characterized by transmittance measurements at different humidity levels in the range from 5% to 90% relative humidity. Additionally, the humidity induced color change is studied by calculating the color coordinates at different relative humidity using measured spectra of transmittance or reflectance. A special attention is paid to the selection of wavelength(s) of measurements and discriminating between different humidity levels when sensing is performed by measuring transmittance at fixed wavelengths. The influence of initial film thickness, sensor architecture, and measuring configuration on sensitivity is studied. The potential and advantages of using top covered Bragg stacks and polymer/metal thin film structures as humidity sensors with simple optical read-outs are demonstrated and discussed.

Keywords: optical sensing; humidity; Bragg stacks; branched polymers

---

1. Introduction

Humidity is present everywhere and is one of the mostly important physical parameters in our lives. The accurate monitoring of humidity is significant in semiconductors, electronics, food processing, and pharmaceutical industries where the quality of products is influenced sufficiently by the humidity. Humidity control is also essential in museums or archives in order to ensure the correct storage of artworks. Further, monitoring of relative humidity in office environments is in favor of human comfort and health and helps achieve hygienic conditions.

The perfect humidity sensor should have high sensitivity, fast response, long-term durability, low cost, easy detection, and wide dynamic range, i.e., it should operate over a wide range of humidity. The established technology nowadays is an electrical measurement of humidity [1] where the change of resistivity, dielectric constant, or the impedance of a sensitive medium, usually metal oxide thin film, is used [2,3].

In recent years, optical sensing of humidity has gained increasing interest and great research efforts have been put in this field [4,5]. Optical sensors operate at room temperature and do not require an electrical power supply, which makes them suitable for flammable and harsh environments and enables operation at high pressures as well. Optical fiber humidity sensors are the most widely spread optical sensors for humidity. Different materials have already been used for their functionalization such as agarose gel [6], graphene oxide [7], polyvinyl alcohol [8], etc. The main disadvantage of this type of sensor is the cost of optical equipment, particularly spectrometers and optical spectrum analyzers.

Cost effective and relatively simpler approaches are color sensing of humidity or monitoring of appropriately chosen optical parameters that change with humidity. In the first approach the detection is performed by visual inspection of color, or color monitoring by an inexpensive camera [9–11]. Different photonic structures have been utilized: reflection holograms [9,10], Bragg stacks [12], and 2D and 3D photonic crystals [11,13]. In the second approach the swelling of a sensitive medium [14] or the transmittance power [15] are measured with humidity.

Polymers attract considerable interest as optical sensing materials due to the advantageous features such as flexibility, light weight, ease of processability, useful mechanical properties, and compatibility with oxides and ceramics [2,3]. In general, natural or synthetic polymers such as hydroxyethyl cellulose, chitosan, poly(methyl methacrylate), polyaniline, poly(vinyl alcohol), and poly(ethylene oxide) are used in optical sensors in the form of films, beads, micro-rods, nano-spheres, etc. [16–18]. Along with the use of single commodity polymers and polymer–inorganic composites, various copolymers [19–22], hybrid structures [23,24], and smart copolymers [25] are synthesized and their optical properties studied in an attempt to address the specific requirements arising from the diverse potential applications.

It is already demonstrated that thin films with submicron thicknesses exhibit faster swelling than bulk hydrogels [26]. We have already shown that in order to achieve a large response in terms of swelling, the polymer thin film must have the right chemistry and the right macromolecular structure [22]. However, the geometry of measurements and the design of the sensor, for example thickness, substrate type, sublayer, top layer, etc., also have considerable impact on sensor sensitivity.

In this paper, for detection of humidity we use transmittance measurements because they are simpler, more accurate, and less expensive compared to reflectance measurements. Moreover, observing sample color in transmittance mode, i.e., using the light transmitted through the sample, could overcome to a great extent the ambiguity related to the angle dependence of the color when reflectance mode is used. We study two sensor configurations. In the first one we deposit the sensitive film on top of the Bragg stack, instead of incorporating it in the stack's backbone, thus overcoming possible issues of incompatibility that may arise when organic and inorganic materials are used for the stack. The second sensor configuration is comprised of a thin metal sublayer deposited on glass substrate and top-covered with the sensitive polymer film. The metal sublayer has dual purposes: to provide color for colorless polymer film and to increase the optical contrast between the polymer and glass substrate that have almost the same refractive indices (around 1.5).

In this study we prove the concept of utilization of top covered Bragg stacks and polymer/metal thin films as optical sensors for humidity. As humidity sensitive media we use thin films of previously developed multiblock copolymer with branched macromolecular architecture comprised of poly(*N*,*N*-dimethyl acrylamide) (PDMA) and poly(ethylene oxide) (PEO) segments [22]. The humidity sensing ability are demonstrated through both transmittance measurements and calculation of CIE (International Commission on Illumination) color coordinates at different relative humidity in the range 5–90% relative humidity (RH). The impact of initial film thickness, sensor architecture, and measuring configuration on sensitivity is studied and discussed.

2. Materials and Methods

The monomer *N*,*N*-dimethylacrylamide (DMA) was purchased from Sigma–Aldrich (Steinheim, Germany) and purified from inhibitor by passing through a column of basic alumina before use. Poly(ethylene oxide) of molar mass 2000 $g \cdot mol^{-1}$ (PEO2000) was supplied from Fluka (Buchs, Switzerland). Initiator diammonium cerium (IV) nitrate $(NH_4)_2Ce(NO_3)_6$ and crosslinker poly(ethylene glycol) diacrylate of reported average Mn 575 (PEGDA) were purchased from Sigma–Aldrich as well. Other solvents and reagents were of standard laboratory reagent grade and used as received.

The synthesis of the copolymer is described in detail in our previous papers [22,27]. Briefly, a redox polymerization of *N,N*-dimethylacrylamide (DMA) in deionized water was carried out using ammonium cerium (IV) nitrate as the initiator. Poly(ethylene oxide) and poly(ethylene glycol) diacrylate were implemented as a hydroxyl functionalized initiating moiety and as a cross-linker, respectively. The polymerization was carried out for 3 hours at 35 °C in nitrogen atmosphere under vigorous stirring and terminated by diluting the reaction mixture with methanol (1:1 volume ratio). The copolymer solution thus obtained was used for thin polymer film deposition without further isolation or purification. In order to control the thickness of the films, the copolymer solution was diluted to the appropriate concentration by using a water/methanol mixture (1:1 volume ratio). Chemical structure of the copolymer is schematically presented in Figure S1.

Thin polymer films with thickness of 100, 200, 300, and 400 nm were deposited by the spin-coating method at a rotation speed of 4000 s^{-1} and duration of 60 s using polymer solutions with concentrations of 4, 6.25, 7.3, and 8.3 g L^{-1}, respectively. A post deposition annealing at 180 °C for 30 min in air was applied to all films. Silicon wafers, Bragg reflectors, and Au–Pd coated glasses were used as substrates. Five- and seven-layer Bragg reflectors were prepared on glass substrates by alternating deposition of sol-gel Nb_2O_5 [28] and SiO_2 [29] films or dense and porous Nb_2O_5 films [30]. The reflectors were designed in a way that their stop bands were in the visible spectral range. The Au–Pd sublayers with thicknesses of 15, 30, and 70 nm were deposited on glass substrates by cathode sputtering of gold/palladium target (Quorum Technologies) for 15, 60, and 90 seconds, respectively, under vacuum 4×10^{-2} mbar using a Mini Sputter Coater SC7620 system. The thickness and diameter of the target were 0.2 and 57 mm, respectively. The purities of gold and palladium were 99.99% and 100%, respectively, and Au:Pd ratio was 80:20. The Au–Pd layers' thickness was determined using mechanical profilometer Talystep (Rank Taylor Hobson, Leicester, UK) and confirmed by 3D optical profiler (Zeta 20, Zeta Instruments, Milpitas, CA, USA).

Refractive index, extinction coefficient, and thickness of polymer films were calculated using previously developed two-stage nonlinear curve fitting of reflectance spectra measured with UV-VIS-NIR spectrophotometer (Cary 5E, Varian, Melbourne, Australia) at normal light incidence [28]. The accuracy is 0.01, 0.005, and 1 nm for the refractive index, extinction coefficient, and thickness, respectively. The sensing behavior was tested by measuring in-situ transmittance or reflectance spectra at different levels of relative humidity (RH), realized using a homemade bubbler system that generates vapors from liquids [31]. The exact values of RH were obtained from a reference humidity sensor placed in the measuring cell. In order to check the applicability of the studied samples for color sensing of humidity, CIE color coordinates were calculated at each humidity level [32] and plotted on X–Y color space. Measured transmittance or reflectance spectra of the samples were used for calculation of CIE color coordinates.

Indeed, the reflectance and transmittance spectra of the obtained thin polymer films were registered in at least three consecutive experiments for each film sample. As the reflectance and transmittance data points derived from the subsequent experiments were practically equal, standard deviations were not added to the figures in the manuscript. The reported accuracy of the spectrophotometer measurement is 0.1% for transmittance and 0.26% for reflectance.

The sensitivity of the sensors S, measured in % per % RH, was calculated according to the following equation:

$$S = \frac{\Delta R}{RH_2 - RH_1} \tag{1}$$

where ΔR (or ΔT, if transmittance T is measured) is the change of sensor's reflectance (or transmittance) in % which is provoked by the alteration of relative humidity from RH_1 to RH_2.

The accuracy/resolution of detection (ΔRH), measured in % RH, was calculated by Equation (2):

$$\Delta RH \;=\; \frac{errR\;(\%)}{S\;(\%/\%RH)} \tag{2}$$

where *errR* (or *errT*, if *T* is measured) is the experimental error (measurement accuracy) of *R* or *T* and *S* is the sensitivity, calculated by Equation (1).

3. Results and Discussion

When polymer film or hydrogel are exposed to high humidity the polymer chains swell due to the penetration of moisture inside the film and as a result they change their thickness, i.e., swelling is observed. Simultaneously, a drop in refractive index takes place due to the reduced density of the film [32]. It may be expected that the degree of swelling depends on the initial thickness of the film. Therefore, the first step of our investigation concerns this dependence. Two possible approaches for conducting this study exist: measuring reflectance or transmittance of the films at different humidity levels or to calculate the change of film thickness with humidity. In our work we choose to monitor the thickness change because film thickness is an intrinsic physical parameter and, unlike transmittance and reflectance, it is independent of other parameters such as refractive index, wavelength, substrate type, etc.

Thin polymer films with different thicknesses (from 100 to 400 nm) were deposited on Si-substrate and exposed sequentially to relative humidity of 5% RH and 70% RH. Reflectance spectra of the films were measured at both humidity levels and optical constants and film thicknesses were determined using non-linear curve fitting [28]. Reflectance spectra of 100 nm and 300 nm thick films are presented in Figure 1a,b, respectively. The increase of the number of interference peaks and decrease of their intensity are clearly seen in the case of thicker films (Figure 1b). The first observation indicates the increase of the thickness of the film when it is exposed to high humidity levels, while the decrease of fringe intensity confirms the decrease of the refractive index associated with the drop of film density that is in correlation with our previous results [32]. The relative thickness increase Δd (in percent), i.e., the swelling, is calculated according to Equation (3):

$$\Delta d \;=\; \left(\frac{d_{70\%} - d_{5\%}}{d_{5\%}} \right) \times 100$$

where $d_{70\%}$ and $d_{5\%}$ are the calculated film thicknesses at relative humidity of 70% RH and 5% RH, respectively.

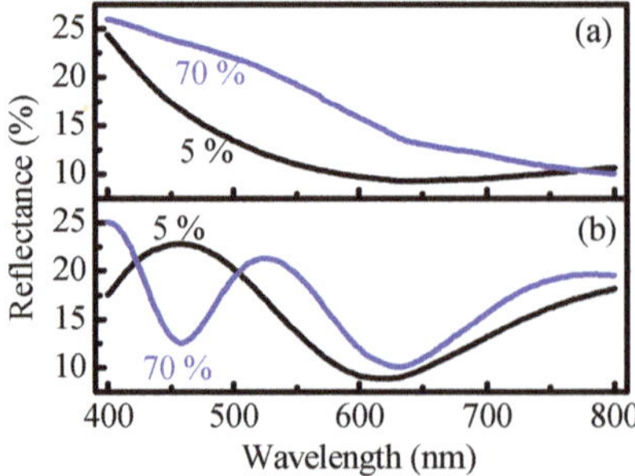

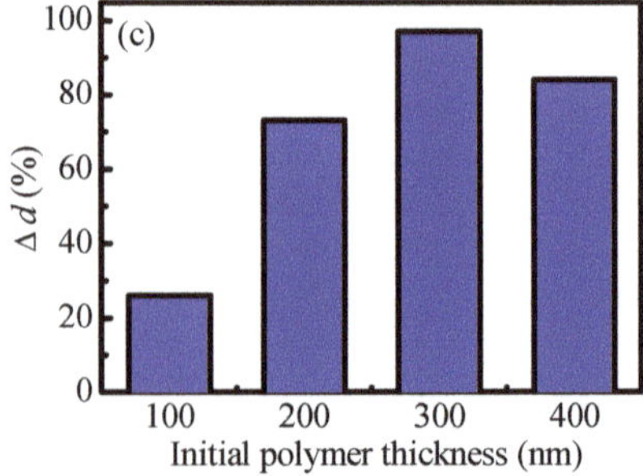

Figure 1. Left: Reflectance spectra at relative humidity (RH) of 5% and 70% of polymer films with initial thickness of 100 nm (**a**) and 300 nm (**b**) deposited on Si substrate; Right: Relative thickness change Δd (in percent) of thin polymer films with initial thicknesses in the range 100–400 nm due to the change of humidity from 5% RH to 70% RH (**c**).

Figure 1c presents relative thickness change as a function of initial polymer film thickness, i.e., the film thickness at 5% RH. It is seen that for all samples the exposure from low to high humidity leads to an increase of the films thickness. However, the degree of swelling depends strongly on the initial thickness of the films and it is the highest for 300 nm thick film. It is seen that Δd increases with thickness: 26%, 73%, and 97% for polymer films with initial thicknesses of 100, 200, and 300 nm, respectively. For the film with initial thickness of 400 nm a small decrease in Δd is observed (84%) compared to those of an initial thickness of 300 nm. The possible reason for the thickness dependence of the swelling is the adhesion to the substrate. It could be expected that the influence of substrate adhesion is the strongest for the thinnest film and decreases gradually with thickness. This explains the fourfold increase in swelling of polymer film with a thickness of 300 nm, as compared to this with a thickness of 100 nm.

In order to determine the response time of both thin (100 nm) and thick (300 nm) polymer films we deposited them on glass substrate and monitored the temporal change of transmittance at fixed wavelength (λ_{max}) when the humidity in the measuring cell goes continuously from 5% RH to 80% RH. Figure 2a presents the temporal change of the humidity in the cell measured by the reference humidity sensor. Figure 2b presents the temporal dynamic of transmittance of thin polymer films with thicknesses of 100 and 300 nm deposited on glass substrate measured at selected wavelengths when the humidity in the cell changed continuously from 5% RH to 80% RH following the dependence presented in Figure 2a. The monitoring wavelength λ_{max} is selected as the wavelength at which the humidity response is the strongest. Considering that the transmittance and reflectance of thin films are non-linear functions of film thickness (Figure S2) it is obvious that λ_{max} depends on the film's optical constants and thickness as well as on the type of substrate used. Glass substrate is selected for this experiment because the oscillatory behavior of the transmittance versus thickness curve is not so strong as compared to the case of the reflectance versus thickness curve when silicon substrate is used (Figure S3).

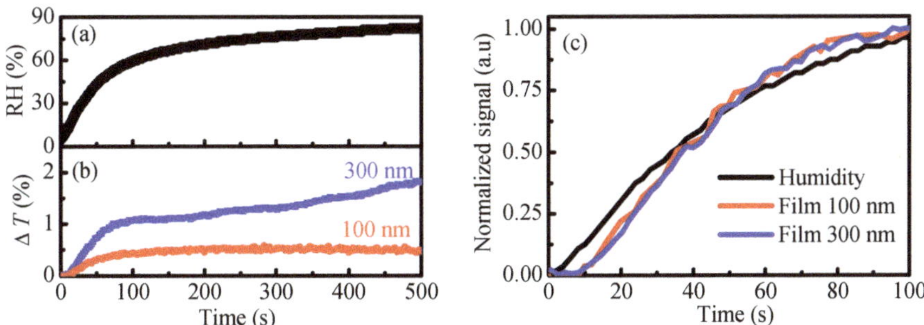

Figure 2. Left: (**a**) Temporal change of the humidity in the cell measured by the reference humidity sensor; (**b**) temporal change of transmittance of thin polymer films with denoted thicknesses deposited on glass substrates when humidity changes gradually from 5% RH to 80% RH; Right: (**c**) Normalized signal for temporal changes of humidity and transmittance of polymer films with thicknesses of 100 and 300 nm.

To study the delay of our sensors relative to the reference sensor, we normalized the measured humidity and transmittance curves and plotted them in Figure 2c. The comparison to a reference sensor shows a delay for low humidity, i.e., at the beginning of measurements that vanishes for RH values greater than 50% RH. For example at 10% RH and 20% RH the delay is 8 and 6 s, respectively, and it decreases to 4 s for 30% RH and 40% RH. Interestingly both polymer films react similarly to the change of humidity and there is no delay for the thicker film in respect to the thinner one. For a rapid change of humidity from 50% RH to 95% RH, achieved by blowing of humid air, the change of color is almost instant. The sensor recovers rapidly when blowing is off.

We have already demonstrated that when humidity is increased, the thickness of all studied polymer films increases as compared to its initial value. Besides, the swelling depends on the initial film thickness and has a maximal value for 300 nm thick film. Further, it is interesting to study how the film thickness changes with humidity, i.e., the humidity dependence of film swelling.

Reflectance spectra at relative humidity of 5, 40, 60, and 90% RH of thin polymer films with thickness of 100, 200, and 300 nm deposited on silicon substrate are presented in Figure 3a–c. For all films substantial change of reflectance spectra with RH is observed. Spectra are further used for calculation of the thickness of the films [28].

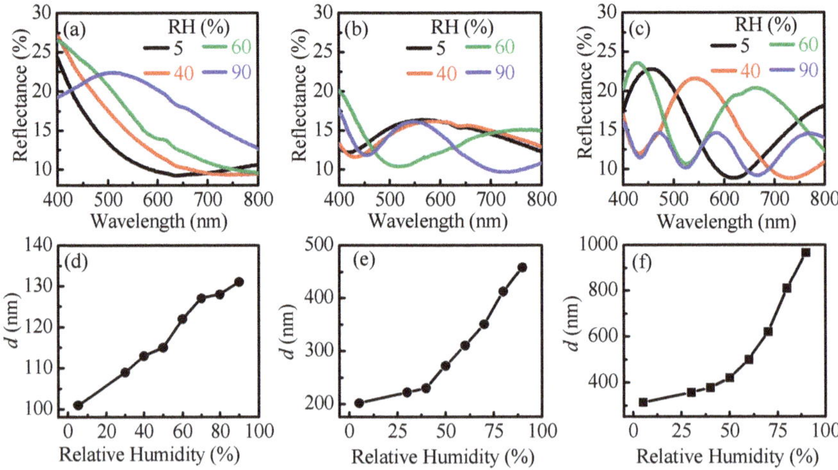

Figure 3. Reflectance spectra at denoted relative humidity values of thin polymer films with thickness of 100 nm (**a**), 200 nm (**b**), and 300 nm (**c**) deposited on silicon substrates. Growth of film thickness with relative humidity for thin polymer films with initial thicknesses of 100 nm (**d**), 200 nm (**e**), and 300 nm (**f**).

The swelling dynamic, i.e., the growth of film thickness with relative humidity, is shown in Figure 3d–f. For the thinnest film an almost linear dependence of *d* versus RH is obtained in the whole studied RH range (5–90% RH). This is beneficial for using the film as a sensing medium for a wide range of humidity. Similar linear dependence is obtained for 200 nm thick film in 40–90% RH. The expansion of the thickest film is exponential with the increase of the relative humidity and changes from 314 nm at 5% RH to 965 nm at 90% RH. The overall change of film thicknesses is 31%, 127%, and 207% for 100, 200, and 300 nm thick films, respectively.

Considering the substantial change in thickness with humidity, we could expect also considerable change of color. If it turns out this is the case, the films could be used as color indicators for humidity. However, first the evolution of color with humidity should be investigated. Using the measured reflectance spectra of films at different humidity we calculated the CIE color coordinates [32] and plotted them in X–Y color space (Figure 4).

The color of the thinnest film is almost the same for RH values in the range from 5% to 40% RH and starts to change gradually from dark blue to light blue for RH higher than 40%. The color coordinates are well separated in the color space thus enabling color sensing of humidity in the range 40–90% RH. Although the thickness values at 5% RH and 90% RH are quite different for thicker films, 202 and 456 nm, respectively, the colors of the films at these humidity levels are very similar (Figure 4b—red and green circles are close to each other). The reason is the periodicity of transmittance and reflectance and respectively of colors with optical thickness of the films (see Figure S2). Remaining points are well separated in X–Y color space (Figure 4b). This means that distinct colors will be observed in a wide humidity range, thus making the 200 nm film suitable for color sensing. When analyzing the color coordinates with humidity for the thickest film (300 nm) (Figure 4c) a conclusion could be made that it is suitable for color sensing in the range 5–60% RH. The sensitivity will be the highest because the separation of the points is the most distinguished and unambiguous in this case.

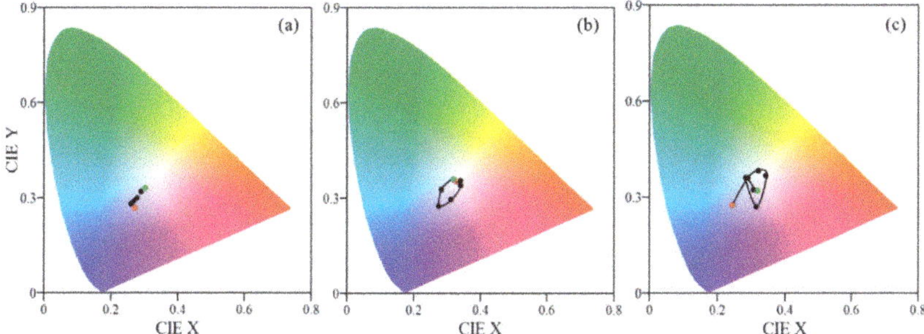

Figure 4. CIE color coordinates for polymer films with initial thickness of 100 nm (**a**), 200 nm (**b**), and 300 nm (**c**) deposited on silicon substrate and exposed to different humidity levels starting from 5% RH (red circle) and ending at 90% RH (green circle). The middle values are 30%, 40%, 50%, 60%, 70%, and 80% RH.

The main drawback when color sensing is used is the dependence of the color on the viewing angle. The observer cannot be sure whether the color change is due to the humidity change or it is because of his position during the color inspection. It is well known that reflectance and transmittance are functions of the incident angle of the light. Because the color perception of the observer is associated with reflectance or transmittance spectra, their change with incident angle of light results in different colors observed at different viewing angles. For example Figure 5 presents pictures of polymer films with different thicknesses deposited on silicon substrate at three viewing angles—10, 30, and 40 degrees. The change of the color with viewing angle for the thickest film is well seen. Fortunately, for thinner films the color changes are not so distinctive. Indeed, for the thinnest one the color change with angle is slightly noticeable.

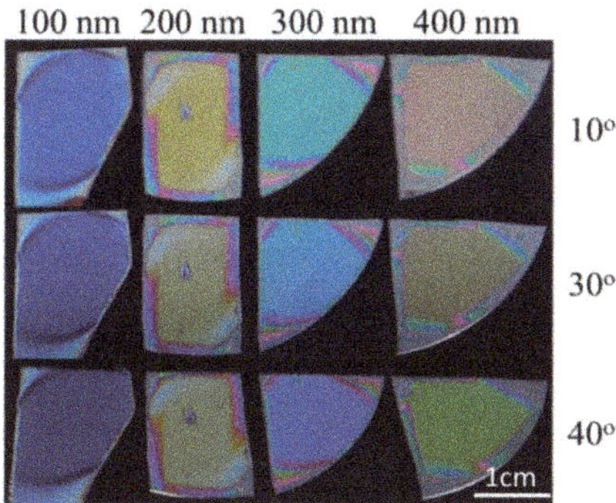

Figure 5. Pictures of polymer films with thicknesses of 100, 200, 300, and 400 nm deposited on silicon substrate at three viewing angles—10, 30, and 40 degrees.

Another possible approach to sensing is monitoring of signal (transmittance, T or reflectance, R) at a fixed wavelength, λ_{max}. As mentioned above, because T and R are nonlinear functions of n, k, d, and λ, the selection of λ_{max} is not a trivial task and has an important role in detection. For example, if we pick a wavelength of 550 nm for λ_{max} for polymer with a thickness of 100 nm (Figure 3a), the measured

R will increase gradually with RH in an approximately linear manner. However, if $\lambda_{max} = 700$ nm than R versus RH dependence will exhibit strong nonlinear behavior.

The humidity dependence of reflectance for different λ_{max} is shown in Figure 6 for all studied samples. For the thinnest one (Figure 6a), the most appropriate λ_{max} is 550 nm. It is seen that at a wavelength of 550 nm the R versus RH dependence is almost linear in the whole humidity range (Figure 6d) and the calculated sensitivity is 0.13% (Equation (1)). If one assumes that the reflectance measurement error is 0.26%, then the lowest humidity step that could be discriminated by the sensor will be 2% RH. Further, if 300 nm thick polymer film is used as sensitive medium and the reflectance of the film is monitored at λ_{max} of 600 nm a twice as high sensitivity of 0.26% could be achieved (the higher incline of the curve R vs. RH in this case is well seen in Figure 6a). This means that the resolution of the sensor will be 1% RH, which is a very good achievement. Unfortunately, the dynamic range of the sensor is limited. This film can be utilized for humidity measurements up to 50% RH.

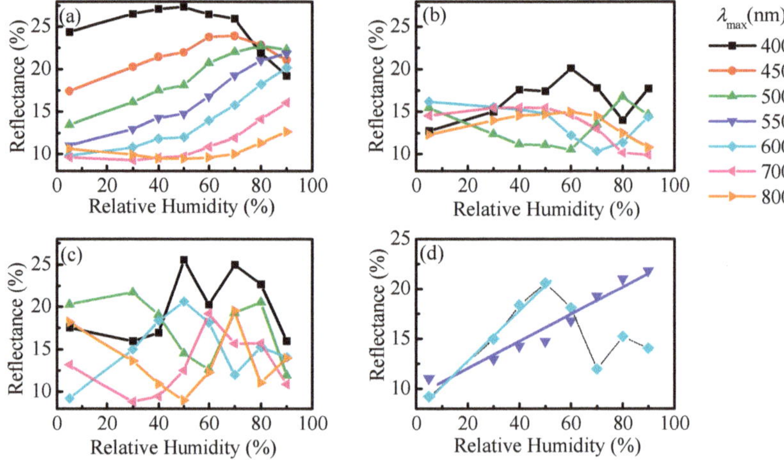

Figure 6. Reflectance at different wavelengths λ_{max} measured as a function of relative humidity for films with initial thicknesses of 100 nm (**a**), 200 nm (**b**), and 300 nm (**c**) deposited on silicon substrates. The best results were obtained for thin film with an initial thickness of 100 nm (blue triangles) and 300 nm (cyan diamonds) (**d**).

From a technological point of view it is more convenient to monitor the transmittance of the film rather than the reflectance. Transmittance measurements are easier to be performed. Besides, they are more accurate and inexpensive. When signal is measured in transmittance mode the angle of incidence could be zero (the so called normal incidence) and the ambiguities associated with angle dependence of color are overcome to the great extent. However, to make transmittance measurements possible, a transparent substrate is required. The cheapest option is to use glass or plastic substrates but in our case they are not suitable because their refractive indexes match the refractive index of polymer thin film and the accuracy drops substantially. Besides, no color will be observed.

Recently we have shown [27,32] that Bragg stacks that exhibit structural colors could be used as transparent substrates. Briefly, Bragg stacks are multilayered systems comprised of alternating layers with low and high refractive indices and quarter-wavelength optical thickness. Because of the quarter-wavelength optical thickness all waves reflected from the layer boundaries in the stack are in phase and interfere constructively. As a result a band of high reflectance appears that is responsible for the observed coloring of the stack. On the contrary, all waves transmitted from different boundaries are out of phase and a band of low transmittance, called a stop band, is generated.

Another advantage of using a Bragg stack as a transparent substrate is illustrated in Figure 7a which presents the pictures of 5- and 7-layer Bragg stacks covered with 300 nm thick polymer films at three different viewing angles. It is seen that the change of the color with viewing angle is weaker as compared to the case of the same film deposited on silicon substrate (Figure 5). This will be beneficial for utilization of polymer top-covered Bragg stacks for color sensing of humidity.

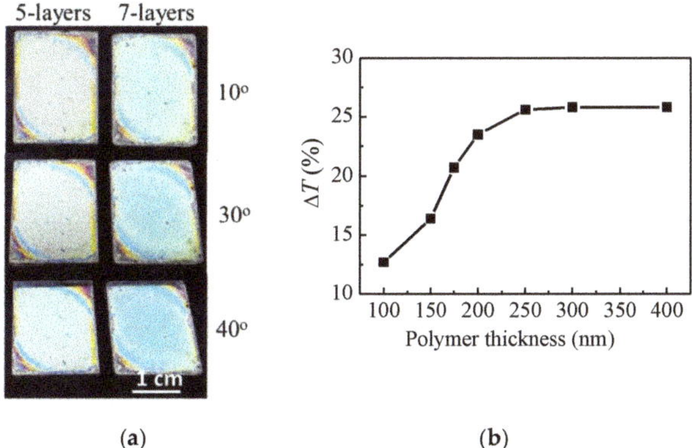

Figure 7. (**a**) Pictures of 5- and 7-layer Bragg stacks covered with polymer film with thicknesses of 300 at three viewing angles—10, 30, and 40 degrees; (**b**) calculated humidity-induced change of transmittance, ΔT, of 5-layer Bragg stacks covered with polymer film with different thicknesses.

In order to achieve the highest sensitivity, we optimized the polymer film thickness through theoretical modeling. The calculated humidity induced change in transmittance (ΔT) of 5-layer Bragg stacks covered with polymer films with different thicknesses in the range 100–400 nm is shown in Figure 7b. It is seen that ΔT increases with the thickness of the polymer film reaching a steady state for thicknesses higher than 250 nm. Therefore, in the next step of our investigation we use polymer films with a thickness of 300 nm and deposit them on four different Bragg stacks comprised of five and seven alternating layers of Nb_2O_5/SiO_2 and dense/mesoporous Nb_2O_5. Our previous results have indicated that the characteristics of stacks, such as operating wavelength, number of layers, optical contrast, etc., do not substantially influence the sensitivity [27]. Still, the best results were obtained for the 5-layer stack consisting of Nb_2O_5 and SiO_2 layers.

Transmittance spectra of the sample at different humidity levels are presented in Figure 8a. In the RH range 5–25% the spectra are almost the same. With increasing humidity, a substantial change is observed. For RH = 30% the shape of the spectra changes; the long-wavelength peak almost disappears, while the short-wavelength one becomes more distinct. With further increase of humidity a shift of the peak toward longer wavelengths takes place.

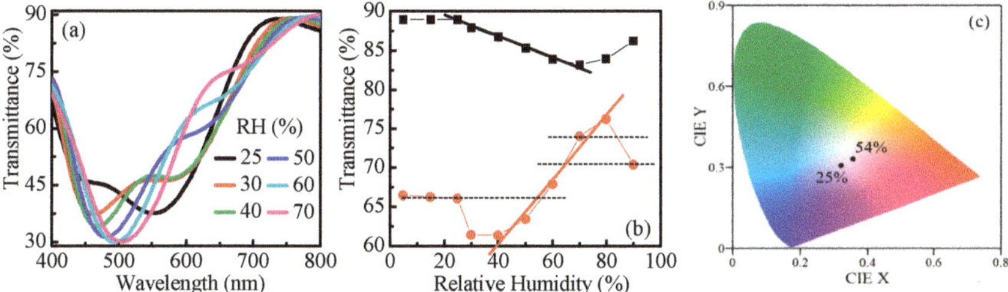

Figure 8. Transmittance spectra (**a**), transmittance values at λ_{max} of 650 nm (red circles) and 730 nm (black squares), (**b**) and calculated CIE color coordinates (**c**) of 5-layer Bragg stacks top-covered with 300 nm polymer film exposed to denoted levels of relative humidity.

The detailed study of the spectra reveals that the most appropriate wavelengths (λ_{max}) for measuring T are 650 and 730 nm (Figure 8b). The obtained sensitivities are 0.37% and 0.13% with dynamic ranges of 40–80% RH and 25–70% RH, respectively. Very good linear dependence of T versus RH is obtained in both cases. Furthermore, the achieved resolutions of detection are very high: 0.3% RH in the 40–80% RH range and 0.8% RH at 25–70% RH. They are calculated with the aid of Equation (2), considering 0.1% experimental error in T.

Although the sensitivity and resolution are high, an ambiguity exists due to the same T-values at different RH (dotted horizontal lines in Figure 8b). For example, the same T will be measured at 25% RH and 54% RH, or at 64% RH and 90% RH, or at 70% RH and 84% RH (the dotted horizontal lines) and the observer will need additional measurements to discriminate between them. We have shown that this could be overcome if color sensing is used simultaneously. Figure 8c presents the calculated color coordinates in transmission mode for relative humidity of 25% and 54%. The points are well separated in the color space thus assuring unambiguous detection of humidity because the observed color of the sample will be different at 25% and 54% and it will not be difficult to discriminate between the two RH levels. The results for other points of ambiguity are very similar and are presented in Figure S4.

Another approach that we consider for humidity sensing by transmittance measurements is implementation of thin metal film deposited on transparent substrate as a transducer. The metal film should be sufficiently thin in order to allow transmittance measurements to be performed. Figure 9a presents transmittance spectra of polymer/metal structures with varying thickness of Au–Pd thin films (15, 30, and 70 nm) and polymer film with a thickness of 300 nm deposited on top. It is seen that T decreases with the thickness of the Au–Pd film from 74% to 33% (the values are at a wavelength of 650 nm). For further experiments we selected Au–Pd film with a thickness of 30 nm because it has an average transmittance of 50%. The change of T spectra with humidity is shown in Figure 9b. As in the case of R of single film on opaque substrate (Figure 3a–c), the number of the interference peaks increases and their intensity decreases, which indicates an increase of thickness and a decrease of refractive index with humidity.

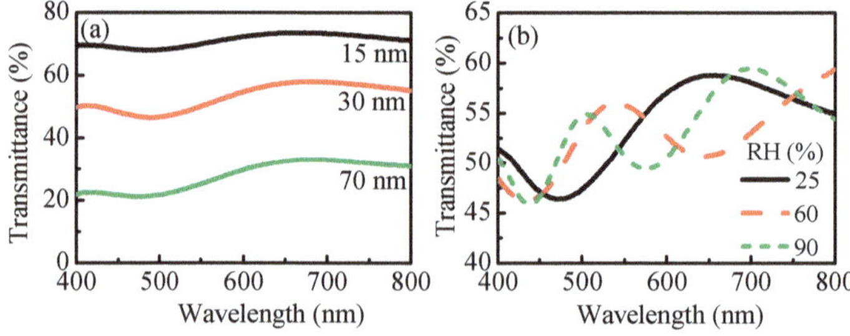

Figure 9. (**a**) Transmittance spectra of Au–Pd films with thicknesses of 15, 30, and 70 nm sputtered on glass substrates and covered with polymer film with a thickness of 300 nm. (**b**) Transmittance spectra of the system polymer (300 nm)/Au–Pd (30 nm)/glass at denoted humidity.

The dependence of T on humidity for the polymer/metal system is shown in Figure 10a. The most appropriate wavelength for measuring T is 700 nm. Two linear ranges in T versus RH dependence are clearly seen. The sensitivity is 0.14% for RH = 25–70% and increases to 0.39% for higher humidity (RH = 70–90%). The respective resolutions are 0.7% RH and 0.3% RH. As in the case above, when using samples comprised of polymer and metal film, ambiguity in the sensing also exists due to the same T-values at different RH. In this case, the discrimination between different RH values is also possible if the color of the sample is monitored simultaneously. Figure 10b,c illustrates the different colors, observed in transmission mode, for the two pairs of RH: (25%, 87%) and (40%, 81%), respectively.

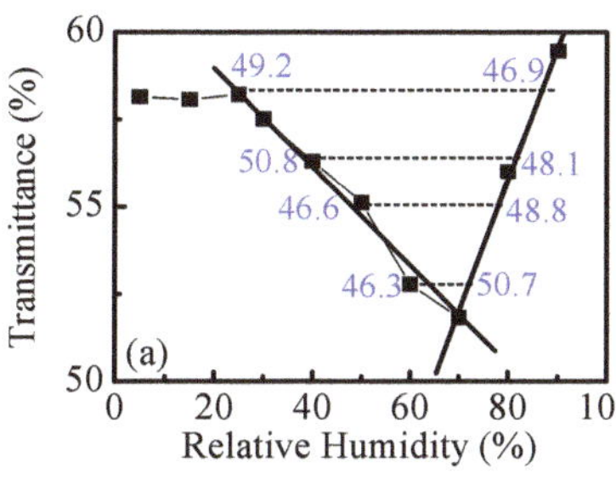
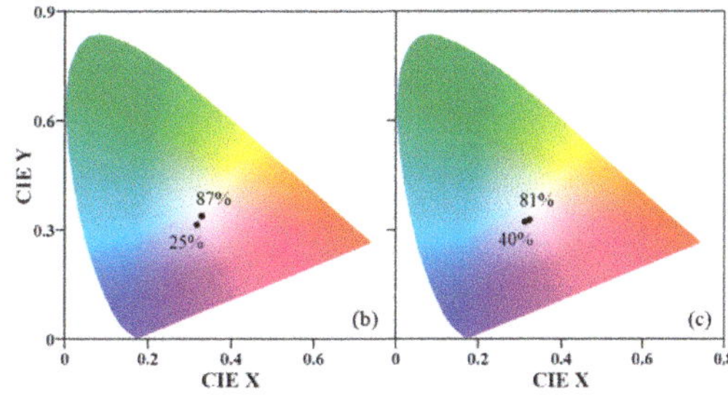

Figure 10. Humidity dependence of transmittance values at λ_{max} = 700 nm (**a**) and calculated CIE color coordinates (**b,c**) of polymer (300 nm)/Au–Pd film (30 nm)/glass exposed to denoted levels of relative humidity. The blue numbers denote transmittance values measured at a wavelength of 430 nm.

Instead of observation of color, the discrimination between RH with the same transmittance values is possible if a second wavelength is used for monitoring of T. As an example, the values of T measured at wavelength of 430 nm are presented as blue numbers in Figure 10a. It is seen that the difference varies between 2.2% and 4.4%, which is sufficient to distinguish different RH exhibiting the same T values at 700 nm.

The obtained results are summarized in Table 1. The widest dynamic range of sensing is obtained for polymer with a thickness of 100 nm deposited on silicon substrate. For sensing, monitoring of reflectance should be applied. Further, it is seen that the implementation of thin metal film as a substrate for 300 nm polymer film is beneficial for increasing the sensitivity and resolution of humidity sensing and enables monitoring of transmittance instead of reflectance. The narrowing of dynamic range is acceptable.

Table 1. Dynamic range, RH, sensitivity, S, and resolution, ΔRH, calculated from Equations (1) and (2), respectively, measured signal type for detection, and sensitive medium.

RH (%)	S (%/% RH)	ΔRH (%)	Signal	Medium
5–90	0.13	2	R at 550 nm	100 nm/Si
5–50	0.26	1	R at 600 nm	300 nm/Si
40–80	0.37	0.3	T at 650 nm	300 nm polymer on Bragg stack
25–70	0.13	0.8	T at 730 nm	300 nm polymer on Bragg stack
25–90	0.14 (25–70) 0.39 (70–90)	0.7 (25–70) 0.3 (70–90)	T at 430 and 700 nm	300 nm polymer on Au/Pd (30 nm)

As mentioned in the Introduction section, the most widely spread and sensitive technology for humidity sensing is fiber optic sensing. Unfortunately they have disadvantages of relatively high cost of optical equipment, particularly spectrometers and optical spectrum analyzers, and complicated preparation of sensitive elements, especially in the case of side-polished optical fibers. The sensitivity of fiber optic sensors is calculated as nm per % RH because usually for detection the wavelength shift of the signal dip is monitored. The sensitivity values reported in the literature vary in a broad range from 0.023 nm/% RH to 1.01 nm/% RH [33]. In order to make a comparison with our sensor, we calculated the resolution using Equation (2) assuming wavelength accuracy of 0.1 nm. The calculated values vary from 4.3% RH to 0.1% RH. Therefore, the value of 0.3% RH achieved by our sensor is very close to the best reported resolution of optical humidity sensing so far. Furthermore, the proposed sensor has the advantages of simple preparation and simple detection. The good sensitivity and resolution along with the technological convenience make the presented sensors very promising and give them important advantages over the widely spread fiber optic sensors.

4. Conclusions

The concept of using polymer top-covered Bragg stacks and polymer/metal thin film structures for optical sensing of humidity was verified and confirmed. The sensing medium was a thin film of branched poly(*N,N*-dimethylacrylamide)-based copolymer with optimized thickness. The sensitivity and resolution of detection as well the dynamic range were compared to the case of single film on silicon substrate. The widest dynamic range of sensing (5–90% RH) was obtained for polymer with a thickness of 100 nm deposited on silicon substrate. The sensitivity was 0.13%, while the resolution was 2% RH. The disadvantage to this method is that, in this case, monitoring of reflectance should be applied. The optimal sensing configuration was a 300 nm thick polymer film deposited on a Au–Pd thin film with a thickness of 30 nm and measuring transmittance signal at two appropriately chosen wavelengths. A sensitivity of 0.14 % was achieved for the 25–70% RH range, which increased to 0.39% for higher humidity (70–90% RH). Relative humidity of 0.7% and 0.3% could be resolved, respectively. There was a narrowing of the dynamic range (25–90% RH), which is acceptable. When a top-covered Bragg stack was used as a humidity sensor a sensitivity of 0.37% and resolution of 0.3% RH could be achieved for a relative humidity range of 40–80% RH.

Supplementary Materials: The following are available online at http://www.mdpi.com/2079-4991/9/6/875/s1, Figure S1: Schematic presentation of the structures of multiblock copolymer with branched macromolecular architecture comprised of poly(N,N-dimethyl acrylamide) (PDMA) and poly(ethylene oxide) (PEO) segments; Figure S2: Calculated transmittance and reflectance at a wavelength of 600 nm of transparent thin film with a refractive index of 1.5 deposited on glass substrate as a function of thickness; Figure S3: Calculated transmittance and reflectance at wavelengths of 600 nm (left) and 800 nm (right) of transparent thin film with a refractive index of 1.5 deposited on glass substrate and silicon wafer as function of thickness; Figure S4: CIE color coordinates of a 5-layer Bragg stack top-covered with 300 nm polymer film exposed to denoted levels of relative humidity.

Author Contributions: Conceptualization, T.B.; Formal analysis, K.L. and T.B.; Investigation, K.L., D.C., R.G., B.G., and T.B.; Writing—original draft, T.B; Writing—review and editing, K.L., D.C., and T.B.

Funding: This research received no external funding.

Acknowledgments: The financial support of the Bulgarian National Science Fund, grant number DN08-15/14.12.2016, is highly appreciated. R. Georgiev and K. Lazarova acknowledge the National Scientific Program for young scientists and postdoctoral fellows, funded by the Bulgarian Ministry of Education and Science with PMC № 577 (17.08.2018). Research equipment from the distributed research infrastructure INFRAMAT (part of the Bulgarian National roadmap for research infrastructures) supported by the Bulgarian Ministry of Education and Science under contract D01-155/28.08.2018 was used in this investigation. The help of Assoc. Prof. K. Lovchinov from IOMT-BAS with reference humidity measurements is highly appreciated.

References

1. Kolpakov, S.; Gordon, N.; Mou, C.; Zhou, K. Toward a New Generation of Photonic Humidity Sensors. *Sensors* **2014**, *14*, 3986–4013. [CrossRef] [PubMed]

2. Sakai, Y.; Sadaoka, Y.; Matsuguchi, M. Humidity sensors based on polymer thin films. *Sens. Actuators B Chem.* **1996**, *35*, 85–90. [CrossRef]

3. Farahani, H.; Wagiran, R.; Hamidon, M.N. Humidity sensors principle, mechanism, and fabrication technologies: A comprehensive review. *Sensors* **2014**, *14*, 7881–7939. [CrossRef] [PubMed]

4. Sikarwar, S.; Yadav, B.C. Opto-electronic humidity sensor: A review. *Sens. Actuators A* **2015**, *233*, 54–70. [CrossRef]

5. Ascorbe, J.; Corres, J.M.; Arregui, F.J.; Matias, I.R. Recent Developments in Fiber Optics Humidity Sensors. *Sensors* **2017**, *17*, 893. [CrossRef] [PubMed]

6. Bariáin, C.; Matías, I.R.; Arregui, F.J.; López-Amo, M. Optical fiber humidity sensor based on a tapered fiber coated with agarose gel. *Sens. Actuators B* **2000**, *69*, 127–131. [CrossRef]

7. Lim, W.H.; Yap, Y.K.; Chong, W.Y.; Ahmad, H. All-Optical Graphene Oxide Humidity Sensors. *Sensors* **2014**, *14*, 24329–24337. [CrossRef] [PubMed]

8. Wang, X.; Zhao, C.L.; Li, J.; Jin, Y.; Ye, M.; Jin, S. Multiplexing of PVA-coated multimode-fiber taper humidity sensors. *Opt. Commun.* **2013**, *308*, 11–14. [CrossRef]

9. Naydenova, I.; Jallapuram, R.; Toal, V.; Martin, S. A visual indication of environmental humidity using a colour changing hologram recorded in a self-developing photopolymer. *Appl. Phys. Lett.* **2008**, *92*, 031109. [CrossRef]

10. Naydenova, I.; Jallapuram, R.; Toal, V.; Martin, S. Characterisation of the humidity and temperature responses of a reflection hologram recorded in acrylamide-based photopolymer. *Sens. Actuators B Chem.* **2009**, *139*, 35–38. [CrossRef]

11. Dalstein, O.; Ceratti, D.R.; Boissière, C.; Grosso, D.; Cattoni, A.; Faustini, M. Nanoimprinted, Submicrometric, MOF-Based 2D Photonic Structures: Toward Easy Selective Vapors Sensing by a Smartphone Camera. *Adv. Funct. Mater.* **2016**, *26*, 81–90. [CrossRef]

12. Lova, P.; Manfredi, G.; Boarino, L.; Comite, A.; Laus, M.; Patrini, M.; Marabelli, F.; Soci, C.; Comoretto, D. Polymer Distributed Bragg Reflectors for Vapor Sensing. *ACS Photonics* **2015**, *2*, 537–543. [CrossRef]

13. Gallego-Gomez, F.; Morales, M.; Blanco, A.; Lopez, C. Bare Silica Opals for Real-Time Humidity Sensing. *Adv. Mater. Technol.* **2018**, 1800493. [CrossRef]

14. Buchberger, A.; Peterka, S.; Coclite, A.; Bergmann, A. Fast Optical Humidity Sensor Based on Hydrogel Thin Film Expansion for Harsh Environment. *Sensors* **2019**, *19*, 999. [CrossRef] [PubMed]

15. Sikarwar, S.; Yadav, B.C.; Dzhardimalieva, G.I.; Golubeva, N.D.; Srivastava, P. Synthesis and characterization of nanostructured MnO_2–CoO and its relevance as an opto-electronic humidity sensing device. *RSC Adv.* **2018**, *8*, 20534–20542. [CrossRef]

16. Lang, C.; Liu, Y.; Cao, K.; Li, Y.; Qu, S. Ultra-compact, fast-responsive and highly-sensitive humidity sensor based on a polymer micro-rod on the end-face of fiber core. *Sens. Actuators B Chem.* **2019**, *290*, 23–27. [CrossRef]

17. Fei, T.; Zhao, H.; Jiang, K.; Zhou, X.; Zhang, T. Polymeric humidity sensors with nonlinear response: Properties and mechanism investigation. *J. Appl. Polym. Sci.* **2013**, *130*, 2056–2061. [CrossRef]

18. D'Amato, R.; Venditti, I.; Russo, M.V.; Falconieri, M. Growth Control and Long range Self-assembly of Polymethylmethacrylate Nanospheres. *J. Appl. Polym. Sci.* **2006**, *102*, 4493–4499. [CrossRef]

19. Venditti, I.; D'Amato, R.; Russo, M.V.; Falconieri, M. Synthesis of conjugated polymeric nanobeads for photonic bandgap materials. *Sens. Actuators B Chem.* **2007**, *126*, 35–40. [CrossRef]

20. Venditti, I.; Fratoddi, I.; Bearzotti, A. Self-assembled copolymeric nanoparticles as chemical interactive materials for humidity sensors. *Nanotechnology* **2010**, *21*, 355502. [CrossRef]

21. Venditti, I.; Fratoddi, I.; Palazzesi, C.; Prosposito, P.; Casalboni, M.; Cametti, C.; Battocchio, C.; Polzonetti, G.; Russo, M.V. Self-assembled nanoparticles of functional copolymers for photonic applications. *J. Colloid Interface Sci.* **2010**, *348*, 424–430. [CrossRef] [PubMed]

22. Lazarova, K.; Vasileva, M.; Ivanova, S.; Novakov, C.; Christova, D.; Babeva, T. Influence of Macromolecular Architecture on the Optical and Humidity-Sensing Properties of Poly(*N,N*-Dimethylacrylamide)-Based Block Copolymers. *Polymers* **2018**, *10*, 769. [CrossRef] [PubMed]

23. Zhao, Q.; Yuan, Z.; Duan, Z.; Jiang, Y.; Li, X.; Li, Z.; Tai, H. An ingenious strategy for improving humidity sensing properties of multi-walled carbon nanotubes via poly-L-lysine modification. *Sens. Actuators B Chem.* **2019**, *289*, 182–185. [CrossRef]

24. Zhang, D.; Xia, J.T.B. Humidity-sensing properties of chemically reduced graphene oxide/polymer nanocomposite film sensor based on layer-by-layer nano self-assembly. *Sens. Actuators B Chem.* **2014**, *197*, 66–72. [CrossRef]

25. Ruiz, J.A.R.; Sanjuán, A.M.; Vallejos, S.; García, F.C.; García, J.M. Smart polymers in micro and nano sensory devices. *Chemosensors* **2018**, *6*, 12. [CrossRef]

26. Stuart, M.A.C.; Huck, W.T.S.; Genzer, J.; Müller, M.; Ober, C.; Stamm, M.; Sukhorukov, G.B.; Szleifer, I.; Tsukruk, V.V.; Urban, M.; et al. Emerging applications of stimuli-responsive polymer materials. *Nat. Mater.* **2010**, *9*, 101–113. [CrossRef] [PubMed]

27. Lazarova, K.; Georgiev, R.; Christova, D.; Babeva, T. Polymer Top-Covered Bragg Reflectors as Optical Humidity Sensors. *Proceedings* **2019**, *3*, 12. [CrossRef]

28. Lazarova, K.; Vasileva, M.; Marinov, G.; Babeva, T. Optical characterization of sol-gel derived Nb_2O_5 thin films. *Opt. Laser Technol.* **2014**, *58*, 114–118. [CrossRef]

29. Lazarova, K.; Awala, H.; El Fallah, J.; Vasileva, M.; Mintova, S.; Babeva, T. Optimization of optical and sensing properties of sol-gel oxides through zeolite doping. *Bulg. Chem. Commun.* **2017**, *49*, 88–94.

30. Georgiev, R.; Christova, D.; Todorova, L.; Georgieva, B.; Vasileva, M.; Babeva, T. Generating porosity in metal oxides thin films through introduction of polymer micelles. *Opt. Quantum Electron.* **2018**, *50*. [CrossRef]

31. Lazarova, K.; Awala, H.; Thomas, S.; Vasileva, M.; Mintova, S.; Babeva, T. Vapor responsive one-dimensional photonic crystals from zeolite nanoparticles and metal oxide films for optical sensing. *Sensors* **2014**, *14*, 12207–12218. [CrossRef] [PubMed]

32. Lazarova, K.; Todorova, L.; Christova, D.; Babeva, T. Color Sensing of Humidity Using Thin Films of Hydrophilic Cationic Copolymers. In Proceedings of the 40th International Spring Seminar on Electronics Technology (ISEE), Sofia, Bulgaria, 10–14 May 2017; pp. 1–6. [CrossRef]

33. Shao, Y.; Wang, Y.; Cao, S.; Huang, Y.; Zhang, L.; Zhang, F.; Liao, C.; Wang, Y. Mechanism and Characteristics of Humidity Sensing with Polyvinyl Alcohol-Coated Fiber Surface Plasmon Resonance Sensor. *Sensors* **2018**, *18*, 2029. [CrossRef] [PubMed]

 *nanomaterials*

Article

Structural and Optical Characteristics of PVA:C-Dot Composites: Tuning the Absorption of Ultra Violet (UV) Region

Shujahadeen B. Aziz [1,2,*], Aso Q. Hassan [3], Sewara J. Mohammed [3], Wrya O. Karim [3], M. F. Z. Kadir [4], H. A. Tajuddin [5] and N. N. M. Y. Chan [5]

[1] Advanced Polymeric Materials Research Laboratory, Department of Physics, College of Science, University of Sulaimani, Qlyasan Street, Sulaimani 46001, Kurdistan Regional Government, Iraq
[2] Komar Research Center (KRC), Komar University of Science and Technology, Sulaimani 46001, Kurdistan Regional Government, Iraq
[3] Department of Chemistry, College of Science, University of Sulaimani, Qlyasan Street, Sulaimani 46001, Kurdistan Regional Government, Iraq; aso.hassan@univsul.edu.iq (A.Q.H.); sewara.mohammed@univsul.edu.iq (S.J.M.); wrya.karim@univsul.edu.iq (W.O.K.)
[4] Centre for Foundation Studies in Science, University of Malaya, Kuala Lumpur 50603, Malaysia; mfzkadir@um.edu.my
[5] Department of Chemistry, College of Science, University of Malaya, Kuala Lumpur 50603, Malaysia; hairul@um.edu.my (H.A.T.); nadianabihahchan@gmail.com (N.N.M.Y.C.)
*   Correspondence: shujahadeenaziz@gmail.com

Received: 16 January 2019; Accepted: 1 February 2019; Published: 6 February 2019

Abstract: In this work the influence of carbon nano-dots (CNDs) on absorption of ultra violet (UV) spectra in hybrid PVA based composites was studied. The FTIR results reveal the complex formation between PVA and CNDs. The shifting was observed in XRD spectrum of PVA:CNDs composites compared to pure PVA. The Debye-Scherrer formula was used to calculate the crystallite size of CNDs and crystalline phases of pure PVA and PVA:CNDs composites. The FESEM images emphasized the presence and dispersion of C-dots on the surface of the composite samples. From the images, a strong and clear absorption was noticed in the spectra. The strong absorption that appeared peaks at 280 nm and 430 nm can be ascribed to the n-π^* and π-π^* transitions, respectively. The absorption edge shifted to lower photon energy sides with increasing CNDs. The luminescence behavior of PVA:CNDs composite was confirmed using digital and photo luminescence (PL) measurements. The optical dielectric constant which is related to the density of states was studied and the optical band gap was characterized accurately using optical dielectric loss parameter. The Taucs model was used to determine the type of electronic transition in the samples.

Keywords: carbon nanodots; hybrid polymer composites; FTIR study; XRD study; optical properties

---

1. Introduction

Since the invention of carbon nano-tubes (CNTs), carbon-based nano-materials have been widely investigated. Carbon quantum dots (CQDs) currently represent the newest class of carbon-based materials as a potential alternative to CNTs for sustainable applications [1]. Carbon nano-dots (CNDs) as a new carbon nano-material with discrete, quasi-spherical carbon nano-particles and ultrafine size of almost 10 nm can be used as a building block for fluorescence systems [2]. Several advantageous characteristics of C-dots, such as an abundance of carbon sources, low cost, biodegradability and brilliant fluorescence behavior, make these new materials widely applicable. Moreover, chemical stability in the colloidal solution state, inertness and relatively high resistivity to photo-bleaching also make C-dots superior over traditional fluorescent organic dyes and semiconductor quantum dots [3]. Recently, has obviously been shown that CNDs have diversity in applications, for instance for biomedical imaging, catalysis, bio-imaging, drug delivery, energy, photovoltaic devices and optoelectronic purposes [2,3]. Because of the abundance of oxygen/hydrogen-containing species such as –OH and –COOH on the surfaces of CNDs, they are chosen as fillers to enhance the hydrogen bonding [4]. Lately, carbon nano-dots (CNDs) have emerged as a new family of light-harvesting materials with remarkable advantages including strong and broad optical absorption, high chemical stability, excellent electron- and hole-transfer capability, and low toxicity [5]. The incorporation of CDs within polymer matrices is under intense study and thus can be utilized in many photonic and optoelectronic applications and integrated in real devices [6]. Organic–inorganic composites are designed for new eras of optical, nonlinear optical, electronic devices and biological labels [7]. As far as we know, polymer nano-composites have attracted the attention of many research groups because of their unique physicochemical properties and wide applications. Moreover, the incorporation of C-dots into solid polymer matrices prevents nano-particles coagulation; as a consequence, higher stability is noticed relative to the colloidal solution counterparts [8]. It is well-known that before applying the organic composites into the devices, their optical properties and morphological profiles have to be characterized. This is due to the fact that any tiny alteration may cause fluctuation in the performance [9]. Herein, poly (vinyl alcohol) (PVA) is considered an attracting artificial polymer that is benign and water soluble, in addition to a relatively high dielectric constant and an excellent film forming capability. All these valuable properties of this polymer can be ascribed to this backbone structure that enables it to form hydrogen bond; as a result, hydrophilic nature dominates and cross linking ability increases with the doping materials [10,11]. In the past decade, immense focus has been devoted to the composites with high transparency and luminescence behavior which have various applications in light emitting devices. Specifically, CD semiconductors are superior to the others in terms of light stability and low toxicity [12]. However, the luminescence-quenching process induced by the particle aggregation limits the application of CDs concerning color tunability and white light fabrication in solid-state illumination systems. The aggregation of CD particles can effectively be avoided via combination with polar polymers [13]. In the current work, a polar, thermo-stable, chemical resistive, easily processible and transparent PVA was used as a hosting polymer in the fabrication of the composite [8]. Recently, extensive research interest has been focused on the improving understanding of the nature of charge transport from the valence to conduction band. This involves attempts to synthesize polymer composites with different ratios between CNDs and the PVA hosting polymer.

2. Synthesis of CDs and Preparation of Polymer Composites

PVA used in this study was supplied by Sigma-Aldrich (Kuala Lumpur, Malaysia). PVA: CNDs polymer nanocomposite films were prepared by the well-known solution casting technique. Hydrothermal treatment of glucose resulted in the formation of the yellow carbon nano-dots (CNDs), as follows: 1 g of glucose was dissolved in 5 mL of concentrated phosphoric acid and the resulting solution was colorless. The solution was then heated in a water bath at (80–90) °C for (20–30) minutes until a dark brown solution was obtained. The solution was cooled down to the room temperature and the pH was adjusted between 3 and 4 using dilute NaOH, afterwards, it was left overnight. The purification of CNDs was conducted using chloroform and then evaporation of the chloroform was performed. The mass of the synthesized CNDs (5 mg) was obtained by subtracting the mass of the beaker from the one of the beakers plus the CNDs. A homogeneous solution of CNDs was obtained by adding 45 mL of distilled water to the CNDs with continuous stirring.

In the preparation of PVA solution, 1 g of PVA was dissolved in 50 mL of distilled water. Afterwards, it was left under continuous stirring at room temperature for 24 hrs until the whole polymer was completely dissolved. As a result, a clear and viscous solution was gained. The final step is preparation of the polymer nano-composite by adding a different portion of CD into a separate container containing PVA solution under a continuous stirring condition. All sample solutions are labeled as CND0, CND1 and CND2 correspond to incorporated PVA solution with 0 mL, 15 mL and 30 mL of 5 mg of dissolved CNDs, respectively. These solutions are further stirred until a homogenous state was achieved. The samples were casted in Petri dishes, and then left for drying at room temperature to allow the film to be formed. The film thickness was controlled in the range of 120–121 μm using constant amount of PVA. Further drying was obtained by transferring the sample solutions into desiccators in an attempt to gain solvent free-films.

3. Characterization Techniques

X-ray diffraction (XRD) data were collected at room temperature using a diffractometer (Bruker AXS, Billerica, MA, USA) operating at a voltage of 40 kV and a current of 40 mA. The samples were scanned with a monochromatic X-radiation beam of wavelength $\lambda = 1.5406$ Å and the glancing angles were in the range of $5° \leq 2\theta \leq 80°$ with a step size of $0.1°$. UV-vis absorption spectra were measured on a Jasco V-570 UV-Vis-NIR spectrophotometer (Jasco SLM-468, Tokyo, Japan) in the absorbance mode. The formation of CNDs-PVA complexes was investigated by Fourier-transform infrared (FTIR) spectroscopy. FTIR spectra were recorded on a Nicolet iS10 FTIR spectrophotometer (Thermo Fischer Scientific, Waltham, MA, USA) in the wave number range of 4000–400 cm^{-1} with a resolution of 2 cm^{-1}. The ATR method was used to measure the FTIR spectrum of the films. The surface morphologies of the PVA:CND composites were examined using Hitachi SU8220 field emission scanning electron microscopy (FESEM) (Europark Fichtenhain A12, 47807 Krefeld, Germany).

4. Results and Discussion

4.1. FTIR Study

FTIR analyses were used to investigate the complex formation in the samples. The FTIR spectra of pure PVA and all the prepared PVA:CND composites are depicted in Figure 1. From the spectrum, the stretching vibration of the hydroxyl groups (O–H) of the pure PVA [14] peaked at 3322 cm^{-1} which increased in intensity and broadness as a result of increasing carbon nano-dots, in CND1 and CND2 doped samples. The wide dispersion of hydrogen bond donor groups, such as –OH and –COOH over the CND surfaces resulted in broadness of the FTIR at 3322 cm^{-1}. This is well-defined that the hydrogen bonding changes both the position and shape of the IR absorption band [4]. Meanwhile, a complex formation between the PVA and CNDs is evidenced from this peak broadening and intensity attenuating. Moreover, the shift of C–H stretching of CH$_2$ group of pure PVA from 2936 cm^{-1} [15], to 2930 and 2931 cm^{-1} was observed in CND1 and CND2, respectively. In addition, there is another shift in C–H bending peak position of pure PVA from 1412 cm^{-1} to 1408 cm^{-1} and 1411 cm^{-1} in CND1 and CND2, respectively. The C–O bending and stretching of the acetyl group on the polymer backbone [16,17] appeared at 1087 cm^{-1} in PVA which also shifted to 1087 cm^{-1} and 1079 cm^{-1} in the respective samples. A vibration peak located between 842 cm^{-1} and 841 cm^{-1} can be ascribed to either C–H rocking mode or C–C stretching [16,18]. Two weak absorption peaks are seen at 1740 cm^{-1} and1370 cm^{-1}, indicating vibrational stretching of C=O and CH$_3$– in acetate moiety which in turn ascribed to incomplete alcoholysis [19]. A relatively sharp peak can clearly been seen at 1561 cm^{-1}, indicating skeletal N–H bending mode [20]. All these changes in the peak positions in the IR spectrum indicate sufficient cross-linking between PVA and CND nano-particles.

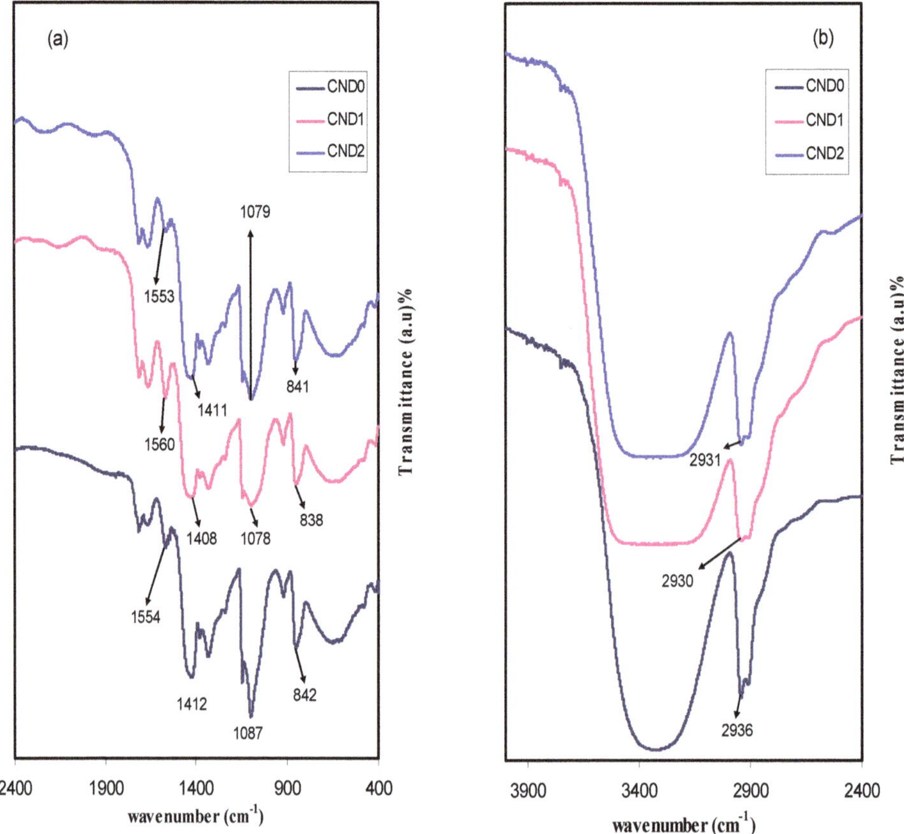

Figure 1. FTIR spectra of all samples in the region of (**a**) 400 cm^{-1} to 2400 cm^{-1}, and (**b**) 2400 cm^{-1} to 4000 cm^{-1}. Clear shifting, broadening and change in intensity in the FTIR bands can be observed.

4.2. XRD and Morphology Study

Figure 2 shows the XRD spectrum of CND particles. It is clear that CDs exhibits a broad crystalline peak at about 2θ = 27.97° and a broad amorphous peak at 42.63°. Previous studies attributed the former peak to highly disordered carbon atoms [21]. Figure 3 represents the XRD pattern of pure PVA and PVA doped with CNDs particles. As can be seen, a broad peak at around 2θ = 20° in pure PVA corresponds to the semi-crystalline nature of the polymer [22]. The XRD pattern (Figure 2) obtained for the C-dots in this work is completely different from that of former work which is relatively broader [23]. For the C-dots obtained, the d-spacing value (0.32) is smaller than that reported (0.34) in the literature [2,23]. It has been proved that the broad peaks in the XRD pattern suggest the nano-scale nature of the prepared particles [24–26]. This can be understood mathematically from the well-known Debye-Scherrer formula:

$$L = K\lambda/\beta\,cos\theta \qquad (1)$$

This means the broader the diffraction peaks are the larger full width at half maximum (β) which in turn led to a smaller crystallite size (L) [27,28]. From Equation (1) the crystallite size was calculated manually using λ = 1.5406 Å, K = 0.9 and the full width at half maximum (β) from the main peak of the XRD pattern at specified 2θ can be estimated. The crystalline size estimated from Equation (1) for the largest peak in the XRD pattern (2θ = 27.97°) of Figure 2 is 1 nm for carbon nano dote (CND) particles.

Therefore, the small size of the C-dots particles is proved from the broad XRD peak. The characteristic feature of C-dots is carbogenic core consisting of both amorphous and crystalline structural parts which enrich in surface functional groups. It is worth-mentioning that in C-dots, amorphous part dominates [2].

The XRD patterns for pure PVA and the PVA doped (CND2) samples (see Figure 3) are evidences for the formation of complexation between PVA and CNDs particles. The relatively large peak centered at 18.6° is shifted to 20.5° for PVA doped. Another two broad peaks appeared at 2θ = 23.4° and 41.18°. From the literature, one can expect that 2θ = 20°, 23.43° and 41.15° belong to (101), (200) and (111) crystalline phases of PVA [19] and these shifts also result from complexation between functional groups of PVA and surface groups of CNDs particles. This study showed that as the concentration of CNDs increased, the intensity decreased and the peaks underwent broadening. These results were caused by the disruption of hydrogen bonding between the surface groups of CNDs and the hydroxyl group in PVA polymer, resulting in it dominating the amorphous part of the composites [14,29]. The calculated crystalline size from equation (1) for peaks 2θ = 18.6° and 20.5° are 6.5 and 4.6 nm in pure PVA (CDN0) and PVA doped one (CND2), respectively. Thus the crystallite size of regular phases or chains of PVA in crystalline regions was reduced upon addition of CNDs particles to PVA. The strong evidences of amorphous domination PVA doped one (CND2) are lowering in intensity and broadening the peaks. The non-existence of peaks for CNDs in PVA doped sample indicates the dissolution of the whole CNDs in the polymer matrix.

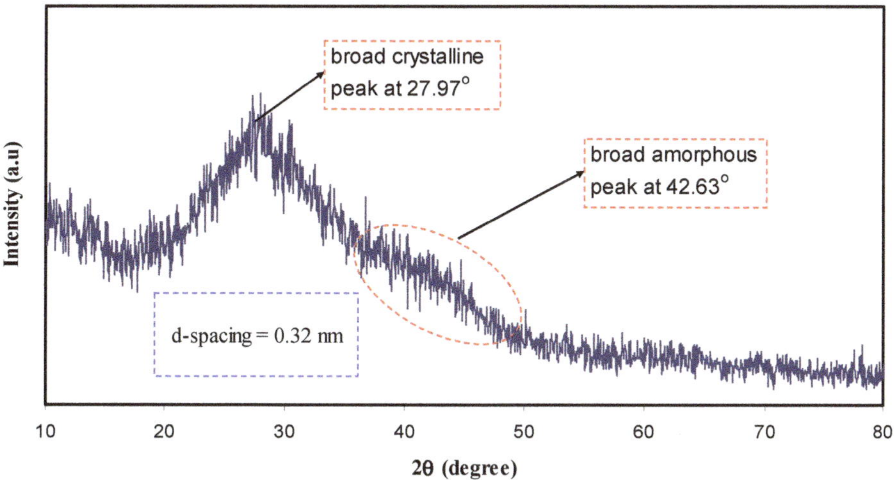

Figure 2. XRD pattern of CN-dots at ambient temperature. Crystalline and amorphous peaks can be seen in the XRD spectrum.

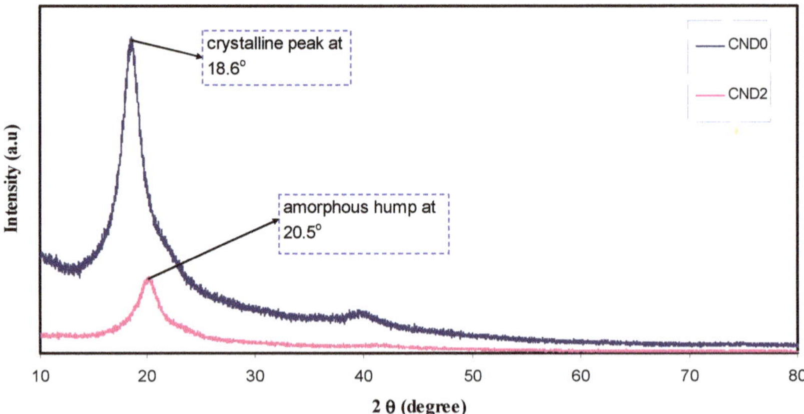

Figure 3. XRD pattern of pure PVA and PVA:CN-Dot composite films. It is interesting to note that the main peak of PVA is more broadened and it intensity decreased after incorporation of CN-dots.

To know the extent of compatibility between the polymer and fillers and also the leakage of the nano-particles to the polymer surface, a commonly utilized technique is field emission scanning electron microscopy (FESEM) [30–32]. The surface morphology was studied using field emission scanning electron microscopy (FESEM). The surface images have shown the formation and distribution of PVA: CND polymer composite. Figure 4a–c show the acquired surface images of both pure PVA and PVA: CND composites, respectively. From the images, one can clearly see the presence and distribution of C-dots on the composite surfaces. Figure 4c exhibits the FESEM image of CND2 sample which incorporated with 30 mL of suspended CND filler. From the image, it is seen that large size CND particles formed on the surface. The FTIR spectra have confirmed the existence of various functional groups, such as –OH and –COOH on the surfaces of CNDs which are incorporated as fillers to enhance hydrogen bonding ability [4]. Another observation that has to be taken into consideration is high density of –OH distribution homogeneously in the polymer [33]. This homogenous distribution infers adequate interfacial interaction between C-dots and polymer matrix as a result of hydrogen bond formation. Therefore, the larger size particles are produced at high concentration. The atomic force microscopy (AFM) has shown the roughness of the surface of PVA:CQD composite at high content of C-dot filler [4].

Figure 4. *Cont.*

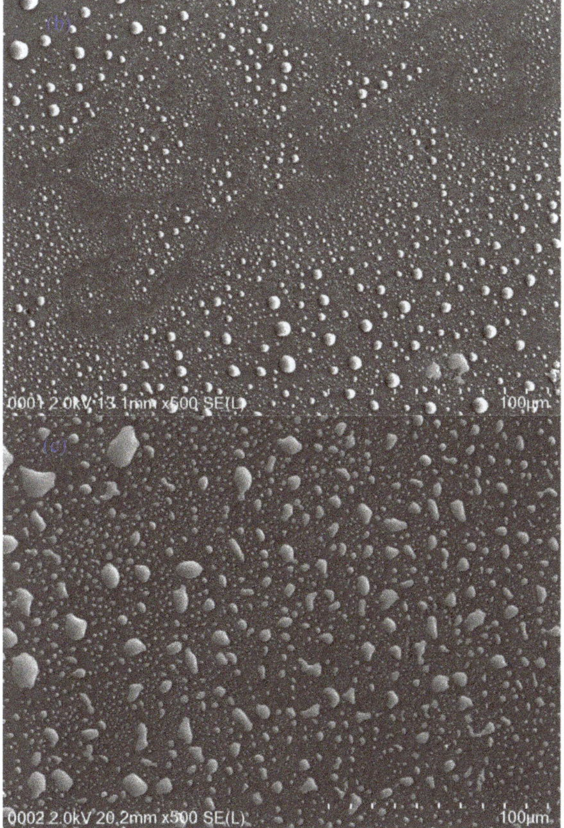

Figure 4. SEM images for (**a**) CND0, (**b**) CND1 and (**c**) CND2 samples.

4.3. Absorption Study

Figure 5 shows the absorption spectra of both pure PVA and PVA:CND film. It is clearly seen that two distinguish peaks have manifested in the UV region. Compared to pure PVA, there is a tuned UV absorption region in the doped ones. The absorption of light in the UV and visible regions result in electron promotion in σ and π and n-orbitals from the ground state to the higher excited state as described by molecular orbital theory. As a consequence, $\sigma \rightarrow \sigma^*$, $n \rightarrow \pi^*$, and $\pi \rightarrow \pi^*$ occur. Most optical transitions are taking place in the visible region caused by impurities. As a result, the generation of defects are color centered [34]. Figure 5 shows detection of two peaks at 280 nm and 430 nm owing to the n-π* and π-π* transitions, respectively [35–37]. Applications in various fields, such as biosensors, imaging probes, viral capsids, QD-based laser, light emitting devices (LEDs) and photovoltaic cells can be ascribed to the unique optical and electronic properties of CDs particles [38]. It is obvious that the onset absorption of the PVA:CND composites is from 580 nm, which lies in the visible region. The lower-mid region of the visible spectrum of the CNDs with their tunable absorption is vital for applications in optoelectronics and sensors to new formulation of bioimaging assays [39]. It is extremely interesting to notice that as the concentration of CNDs increases, the absorption intensities of n-π* and π-π* transitions are also increased. This can strongly be related to the high density of both –OH and –NH$_2$ groups on the CNDs surface [4,9,37]. To check the fluorescence behavior of the PVA:CNDs composite samples, a UV lamp was used in a dark box. Figure 6 exhibits a digital photograph of apparent yellow luminescence of the composite under UV exposure.

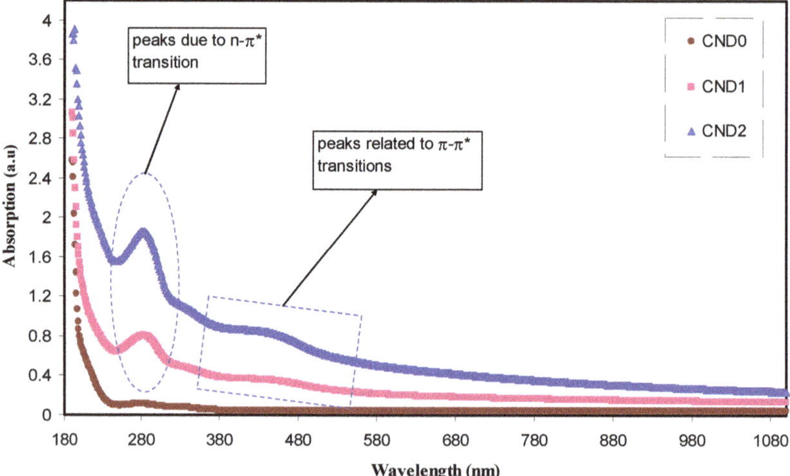

Figure 5. Absorption spectra for all the films. Clearly with increasing CNDs concentration the absorption shifts to higher wavelengths.

Figure 6. The digital photograph of CND2 sample, under UV light exposure.

UV-vis is an informative technique for studying the electronic transitions. Band strength or band-gap energy can be measured from absorption edge in crystalline and non-crystalline materials [40]. The absorption edge is a region in which an electron is jumped from a lower energy state to a higher energy state by an incident photon. The following equation was used to calculate the optical absorption coefficient from the transmittance and reflectance spectra of the films [41]:

$$\alpha = \frac{1}{t} Ln\left(\frac{T}{(1-R)^2}\right) \tag{2}$$

where t, T and R are the thickness, transmittance and reflectance of the sample, respectively. Gradually increasing the absorption coefficient with applied photon energies indicates the amorphous nature of the samples [41]. Figure 7 presents the absorption coefficient as a function of photon energy for pure PVA and PVA:CNDs films. A clear red shift from 6.2 eV to 5.3 eV corresponds to the absorption edge. Clearly, with an increase in CNDs concentration the absorption edge shifts towards lower photon energy. The shift in absorption edge might have resulted from the formation of conjugated bond system caused by bond cleavage and reconstruction. This supports the structural and chemical modifications of PVA incorporated with CNDs particles [42].

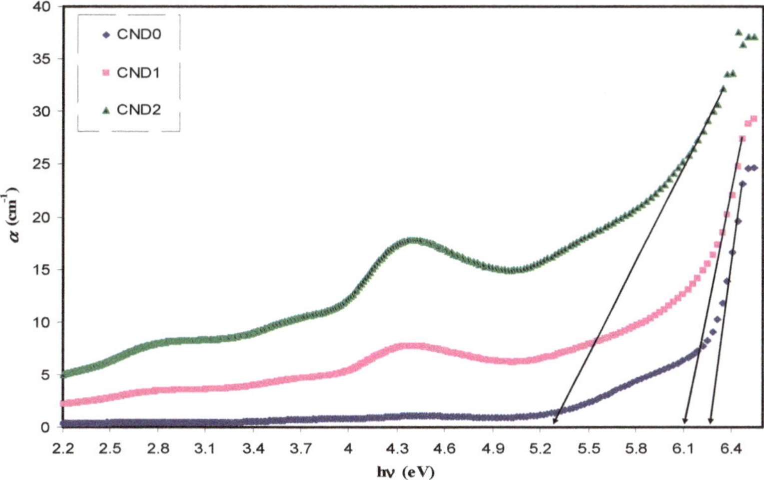

Figure 7. Absorption coefficient versus photon energy for all the films. Clearly with increasing CNDs concentration the absorption edge shifts lower photon energy.

Figure 8 shows the PL spectra for the PVA:CND composite samples at excitation wavelength of 431 nm. Recently, a great deal of research has been devoted to PL of C-dots which is one of characteristics of C-dots that applied in the photocatalysis. The Stokes type emission is obeyed by PL emission, which has a longer wavelength than the excitation one [2]. This emission formed after absorption of photons (electromagnetic radiation). The relaxing and cooling of carrier distribution led to a decrease in width of the PL peak and also to emission energy shifts to match the ground state of the exaction. As the carrier density is increased, the appearance of additional peaks from higher sub-band transitions occur and an increase in the excitation density changes the whole emission spectra [43]. It is obviously noted that as the concentration of C-dot increased, the PL peak intensity increases. This UV absorption resulted in the appearance of characteristic peaks in PL spectra. It is well-known that the excitation UV has shorter wavelength than the emission. It is documented that emission of 385 nm is resulted from the extended conjugation of π-electron domains (island) present in C-dots. The presence of surface trap states (STS) in C-dots led to a strong emission at 460 nm. Recently, C-dots have been characteristic with the existence of STS that contributes in electronic conduction. Clearly, accommodation of electrons in STS facilitates emission at 460 nm; as a consequence, it determines the electronic nature of C-dots [1]. It is well-defined that most PL emission observations can be classified to some extent into two main categories; firstly, one is owing to band gap transitions resulted from conjugated π-domains and secondly, it is due to the defects in the graphene structures. The two factors are synergic in many cases. More clearly, the exploitation and manipulation of defects in graphene sheets results in the creation or induction of the p-domains [2].

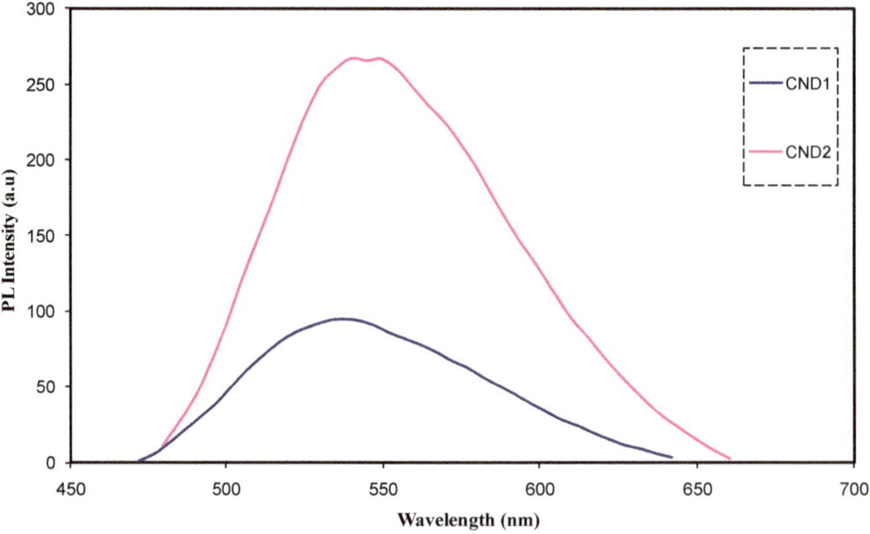

Figure 8. PL spectra for PVA:CNDs composite films at excitation wavelength of 431 nm.

4.4. Refractive Index and Optical Dielectric Constant Studies

Today, studies on the electrical and optical properties of polymers have shown a great deal in view of their applications in optical devices with remarkable reflection, antireflection, interference and polarization properties. The optical properties of polymers can be properly modified by the addition of dopants depending on their reactivity with the host matrix [26,44,45]. One of the parameters is the optical refractive index (*n*), which is the measure of the reduction rate of the speed of light in the medium. The refractive index of the samples has been calculated from the reflectance (*R*) and extinction coefficient (*K*) by using the following equation [26],

$$n = \left[\frac{(1+R)}{(1-R)} \right] + \sqrt{\frac{4 \times R}{(1-R)^2} - K^2} \tag{3}$$

From the following mathematical relationship:

$K = \alpha\lambda/4\pi t$

The extinction coefficient, *K* is directly proportional to both absorption coefficient (*α*) and *λ* is the wavelength whereas inversely proportional to the sample thickness (*t*).

In another mathematical expression, one can clearly see that the reflectance (*R*) can be computed from the absorption (*A*) and transmittance (*T*) values (*R = 1−(A + T)*). The *T* values are calculated from Beer's law (*T = 10^−A*).

From Figure 9, the refractive index spectra of pure PVA and the doped samples are shown. It is apparently revealed that there is a direct proportionality between refractive index and CND concentration. The value of the refractive index is greater than one because of slowing down of photons as a result of interaction with electrons of the host material. It is well-known that the phase velocity of light in vacuum is c = 2.99 × 10⁸ m/s equals the group velocity which is independent of the optical frequency. In contrast, it is typically smaller by a factor n, called the refractive index, in a medium which is frequency dependent. Thereby, the higher value of refractive index of PVA:CND composite the more slowing down occurs of the phase velocity. This is due to the fact that film incorporated with C-dots results in an increase in density and as a consequence, the refractive index becoming increasing predictable [46]. This is also relevant to fact that the refractive index is a function of density, which in other words, is related to the polarizability of the medium [47]. Moreover, for the all doped films, the dispersion of the refractive index versus wavelength is observed compared to the pure PVA. This dispersion behavior of *n* in the doped samples can be related to the increase of density. For additional interpretation, two typical peaks are seen in the spectra of refractive index of the composite samples which can be correlated to aromatic π-π* and n-π* transitions, corresponding to C=C, C=O respectively [48]. It is worth noticing that as the CND concentration increases, the peak intensity increases as a result of populations of more electrons and the number of surface groups. In an attempt to estimate the refractive index of the films, the long wavelength region was extended to Y-axis.

Figure 10 shows the plot between refractive index and CND concentrations. This linearity in the plot is reported in several studies and considered as a satisfactory dispersion of fillers throughout the polymer matrix [18,49–52]. In the data analysis, the r^2 value was determined to be 0.99 from the fitted regression line. This indicated the homogeneity of dispersion of CND throughout the PVA polymer. Furthermore, it is proved to large extend that the refractive index is related to optical dielectric constant (ε_1) parameter which directly related to the localization of electronic states within the forbidden gap of materials [18,22,53]:

$$\varepsilon_1 = n^2 - K^2 = \varepsilon_\infty - \frac{e^2}{4\pi C^2 \varepsilon_o}\frac{N}{m^*}\lambda^2 \tag{4}$$

where ε_∞ and ε_o are the dielectric constant at higher wavelengths and the free space dielectric constant, respectively. N/m^* is the ratio of localized electronic state density to the effective mass, K is the extinction coefficient and e and C have their usual meanings. Figure 11 exhibits the variation of the optical dielectric constant (ε_1) with wavelength at different CNDs concentrations. The ε_1 value and CND concentration have direct proportionality. An increase in ε_1 value from 1.3 to 2.4 can be ascribed to the increment of the density of states because of direct correlation of ε_1 parameter to the density of states inside the forbidden gap of the solid polymer films [22]. The relationship between the static dielectric constant (ε_0) within long wavelengths is well reported [54]. Accordingly, Penn model explains the optical dielectric constant that can strongly be correlated with optical band gap (E_0) [55] as follows:

$$\varepsilon_{(o)} \approx 1 + (\hbar \omega_p / E_o)^2 \tag{5}$$

However, this model has involved the refractive index (*n*) considering $\varepsilon = n^2$. Thus, the Penn model should be expressed on the basis of the refractive index [56].

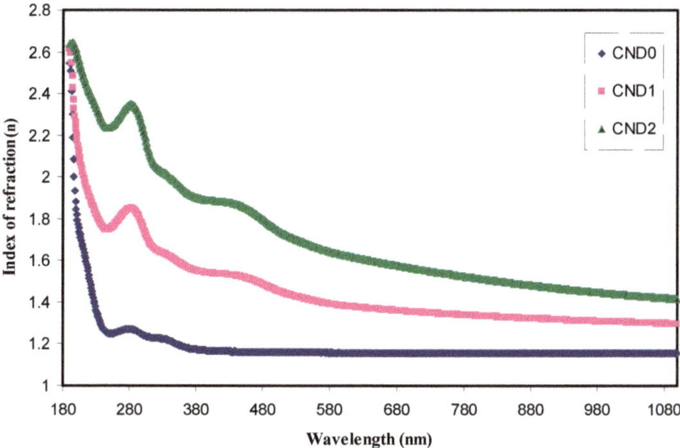

Figure 9. The index of refraction versus wavelength for all the films. Clearly with increasing CNDs concentration the dispersion increased.

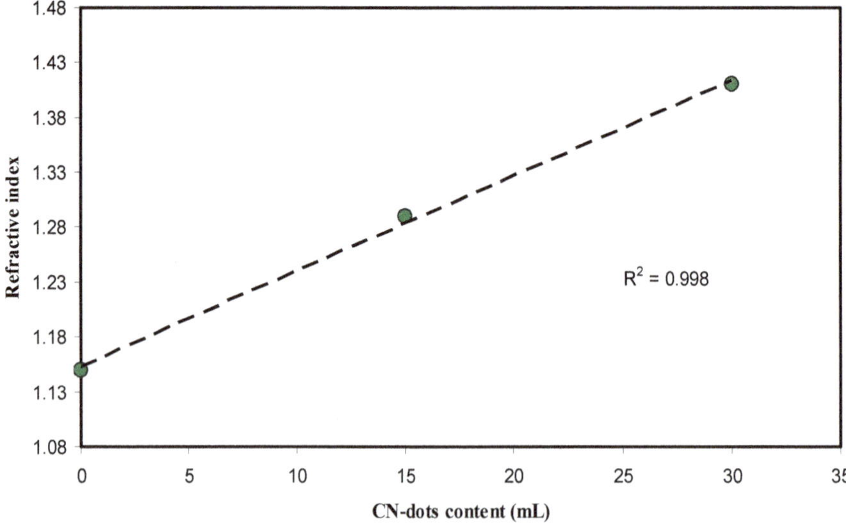

Figure 10. The index of refraction versus CNDs concentration. The linear increase reveals the homogeneous dispersion of CNDs particles.

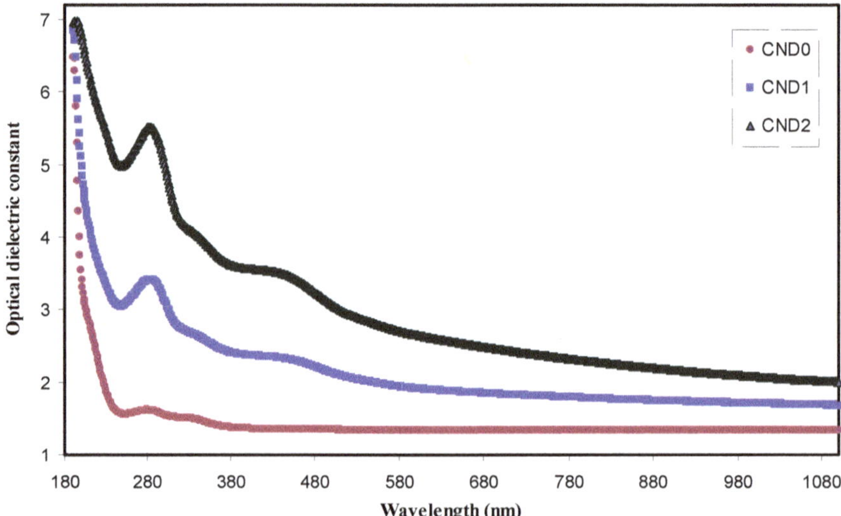

Figure 11. The optical dielectric constant versus wavelength for all the films. Clearly with increasing CNDs concentration the ε' increased.

4.5. Bandgap Study

An Interband absorption process deals with transition of electrons between the bands of solid materials. The absorption edge originated from the onset of optical transitions across the fundamental band gap [57]. Our recent achievements revealed that the fundamental absorption edge derived from the dielectric loss represents the energy bandgap [18,41,44,51,52,58]. These are supported by quantum methods for bandgap investigations. Knowledge of both real and imaginary parts of the dielectric function allows calculating important optical functions [59]. The following complex optical dielectric function is usually used to describe the optical properties of a solid material, interrelated with photon and electron interactions [60].

$$\varepsilon^* = \varepsilon_{1(\omega)} + j\varepsilon_{2(\omega)} \tag{6}$$

$\varepsilon_{1(\omega)}$ as real part and $\varepsilon_{2(\omega)}$ as imaginary part of the complex are associated with electronic polarizability and electronic absorption of the material, respectively [61]. To better understand the electronic structure of different materials, the optical functions have to be studied. Frequency-dependent dielectric function is closely linked to the electronic band structure. Accordingly, the optical properties of homogeneous mediums at all photon energies can be characterized [62]. From the quantum mechanical aspect, the transitions between occupied and unoccupied states are highly related to optical dielectric loss parameter [18,41,44,51,52,58,62]. The equation below can directly be used to determine the imaginary part $\varepsilon_{2(\omega)}$ of the complex dielectric function from the momentum matrix elements between the occupied and the unoccupied electronic states.

$$\varepsilon_2 = \frac{2\pi e^2}{\Omega \varepsilon_o} \sum_{v,c,k} \left| \left\langle \Psi_k^c \left| \vec{u}\,\vec{r} \right| \Psi_k^v \right\rangle \right|^2 \delta(E_k^c - E_k^v - \hbar\omega) \tag{7}$$

where ω is the frequency of light, e the electronic charge, $\vec{u}$ the vector defining the polarization of the incident electric field, and (E_k^c) and (E_k^v) the conduction and valence band wave functions at k, respectively [62]. It is well documented that the fundamental absorption edge in optical dielectric loss spectra provides the optical band gap [63]. A rapid rise near the absorption edge can be a direct evidence for band gap determination [64,65]. The main contributions to the optical spectra are derived from the top valence band (VB) to the lower conduction bands (CB), known as the fundamental absorption edge. The critical points obtained are associated with the band-gap values [60]. Metallic, semiconducting or insulating characteristics of a material can be determined from the electronic properties [66]. Figure 12 reveals the plot between the optical dielectric loss and photon energy for all the samples. The intersection of linear part of ε_2 with the photon energy axis was exploited to estimate the optical band gaps. The estimated values from the plot are listed in Table 1. At a first glance, the higher CNDs concentration, the band gap is reduced. From previous studies, it was concluded that the polymer composites with reduced optical band gap are crucial for photovoltaic and other optoelectronic applications. Huang et. al., reported that the power conversion efficiency of almost 12% of polymer-fullerene-based bulk hetero-junction solar cell incorporated with C-dot particles is increased as a consequence of effective light conversion of near ultraviolet and blue-violet portions of sunlight [67]. Earlier study documented that C-Dot/polymer composites showed relatively acceptable photo-stability, and thus can be used as environmentally friendly composite, low-cost phosphors for solid-state lighting and wave guide applications [68]. Another important aspect in band gap study is the specification of the type of electronic transition. When a photon with a sufficient energy absorbed by an electron transition occurs from the top of the valence band to the bottom of the conduction band and the electron transition obeys some quantum mechanical rules. Tauc's method was applied to specify the type of electronic transition. Optical absorption spectrum is significant for studying the physical properties of polymers comprising the study of a band construction and electronic properties when at pure and doped states [69]. In fact, the optical band gap was estimated from the data of absorption coefficient versus wavelength using Tauc's equation.

$$\alpha h\upsilon = B(h\upsilon - E_g)^n \tag{8}$$

where α, $h\upsilon$, B and E_g denote the absorption coefficient, the photon energy, the band form parameter and the optical bandgap of the samples, respectively, and n a constant, being related to the density of states distribution and determined the type of transition. The n values are equal to 1/2 and 3/2 for direct allowed and forbidden transitions, respectively, while n values of 2 and 3 are for the case of indirect allowed and forbidden transitions, respectively [58]. A direct transition proceeds when the electrons wave vector remains unchanged. However, the interaction with a lattice vibration occurs in the indirect transition, in which the lowest region of the CB locates at different part of the k-space from the highest region of the VB [69]. The plot of $(\alpha h v)^{1/n}$ versus photon energy (hv) when n = 3/2, 2 and 3 is shown in Figures 13–15. In the data analysis, from these figures, various optical band gaps were estimated and tabulated in Table 1. The results showed that the band gaps of the composite films are reduced compared to pure PVA, as a consequence of insulating properties of PVA. Variations in the band gap values make hard to identify a dominant type of electronic transition. In order to specify the exact type of electronic transition in the samples, the band gaps obtained from Tauc's method (i.e., Figures 13–15) were compared to the optical band gaps derived from optical dielectric loss plot (i.e., Figure 12). As a result of comparison, the type of forbidden direct transition (i.e., n = 3/2) can be deduced. To determine the band gap and understand the electron transition phenomena from the top of VB to the bottom of CB both optical dielectric loss parameter and Tauc's model have to be tested. In our previous works, the use of optical dielectric loss parameter for studying the bandgap has been well established [18,41,44,51,52,58,69]. Thus, the present work is an additional support to our hypothesis implying that optical dielectric loss parameter and Tauc's model are sufficient to study the band gap and electron transition types, respectively. From Table 1 it is clear that the band gap decreased upon the addition of CNDs to PVA. A decrease in E_g upon increasing CND concentration might result from the creation of a higher number of free charge carriers/radicals [42]. From the results, one can conclude that these samples can be used for various optoelectronics and previous studies revealed the use of carbon particles in various applications. Huang et. al. [70], used the hybrid materials of carbon and graphene for the fabrication flexible strain sensors and Ke et. al. [71], used carbon dots as fluorescent sensors. Moreover, the capacitance value of supercapacitors is increased upon incorporation of carbon dots to the electrode materials [72]. Thus, CNDs particles have a wide application from optoelectronics to electrochemical devices.

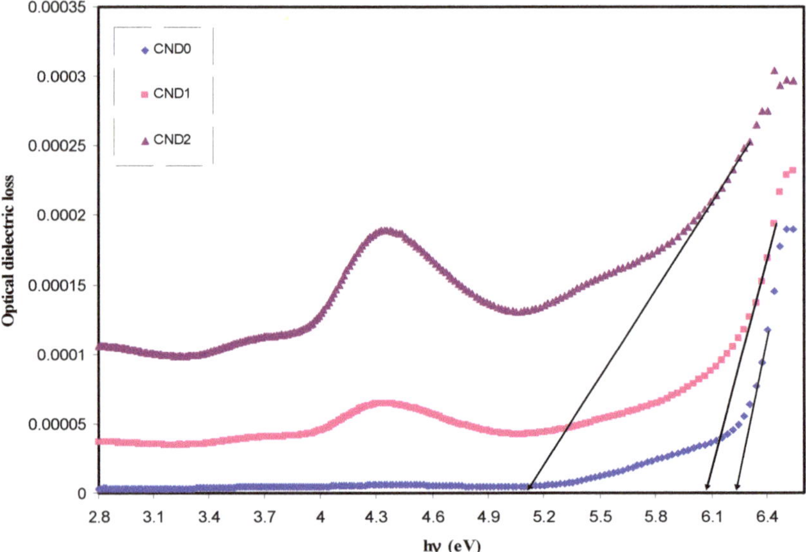

Figure 12. Optical dielectric loss versus photon energy (hv) for all samples. Distinguishable linear parts can be manifested at high photon energy region which can be used to estimate the optical band gap.

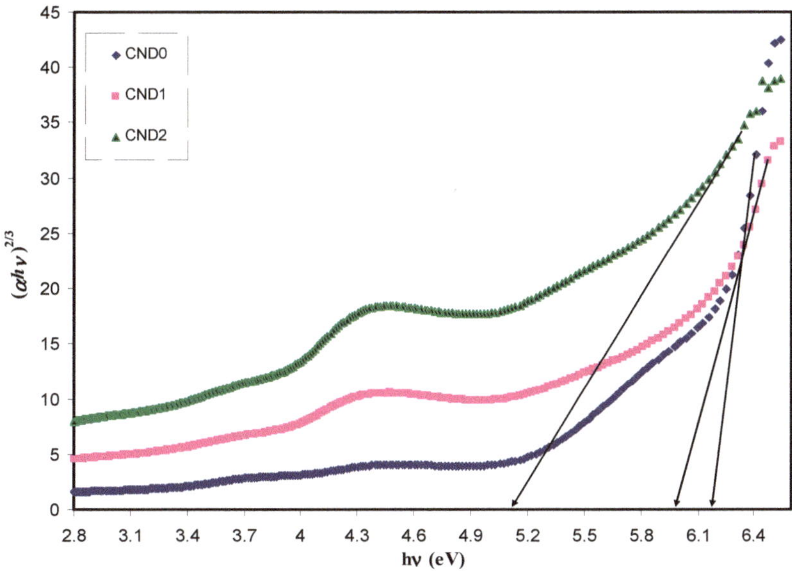

Figure 13. Plot of $(\alpha h v)^{2/3}$ versus photon energy (hv) for all the samples.

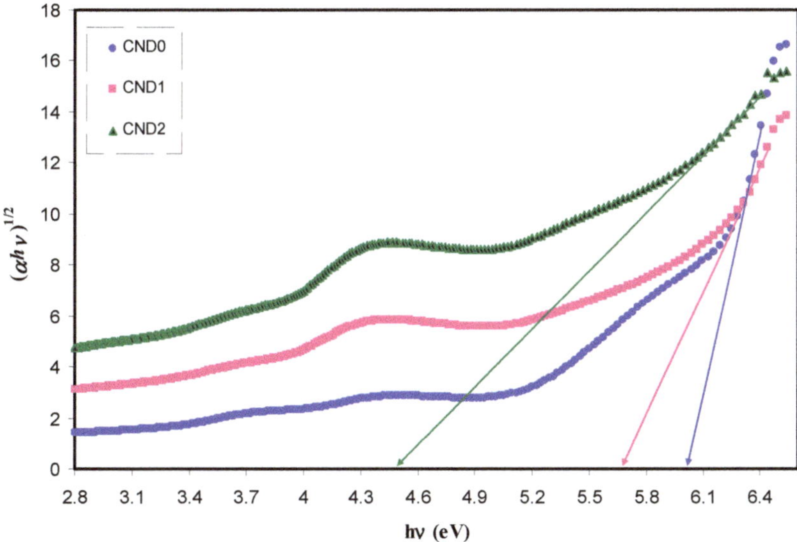

Figure 14. Plot of $(\alpha h v)^{1/3}$ versus photon energy (hv) for all the samples.

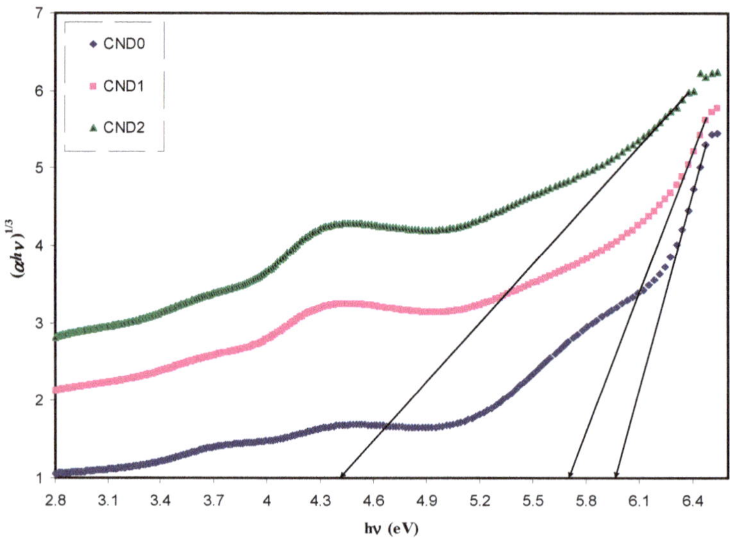

Figure 15. The plots of $(\alpha h\upsilon)^{1/3}$ vs $(h\upsilon)$ for all the samples.

Table 1. Estimated bandgap from Taucs model and optical dielectric loss plots.

Sample Designation	E_g(eV) from Tauc Method (n = 3)	E_g(eV) from Tauc Method (n = 2)	E_g (eV) from Tauc Method (n = 3/2)	Estimated Bandgap from ε''-hυ Plot
CND0	5.95	6	6.18	6.2
CND1	5.7	5.67	5.96	6.05
CND2	4.4	4.5	5.12	5.12

5. Conclusions

In conclusions, fabrication of PVA:CNDs is fascinating after characterizations using a range of spectroscopic techniques. Changes in the FTIR spectral features indicated a good cross-linking between PVA and CND nano-particles. Shifts in XRD spectra of the composites confirmed the complex formation between them. The Debye-Scherrer formula was used to calculate the crystallite size of CNDs and crystalline area of pure PVA and PVA:CNDs composites. The crystallite size of PVA was reduced upon the addition of CNDs particles. Furthermore, the effect of these nano-particles on tuning the absorption of UV spectra in the nano-composites was studied. Strong absorptions at 280 and 330 nm were assigned to n-π^* and π-π^* transition. A reduction in the optical band gap resulted from a shift in absorption edge to lower photon energy. The existence and dispersion of C-dots with different sizes on the surface of composites films were proved using FESEM images. The luminescence behavior of PVA:CND composites was verified using digital photograph and photo luminescence (PL) measurement and the PL intensity increased with an increase in CNDs concentrations. A linear increase of the refractive index with increasing CND concentration revealed a homogeneous distribution of the particles throughout the host PVA. Optical dielectric loss parameter was utilized to estimate the optical band gap. The study of Tauc's model established a forbidden direct type of electronic transition. The results of the present work are an additional support to our hypothesis implying that optical dielectric loss parameter and Tauc's model are sufficient to study the bandgap and electron transition types, respectively.

Author Contributions: Conceptualization, Writing—Original Draft Preparation S.B.A.; Methodology, Investigation A.Q.H., S.J.M., H.A.T. and N.N.M.Y.C.; Writing—Review & Editing, W.O.K and M.F.Z.K.; formal analysis S.B.A., A.Q.H., and S.J.M.; Validation S.B.A. and M.F.Z.K.; Project administration, S.B.A.

Funding: This research was funded by Kurdistan National Research Council, (KNRC), Ministry of Higher Education and Scientific Research-KRG/Iraq. The APC was funded by authors.

Acknowledgments: The authors gratefully acknowledge the financial support for this study from the Department of Physics, College of Science, University of Sulaimani, Sulaimani, and Komar Research Center (KRC), Komar University of Science and Technology, Sulaimani, 46001, Kurdistan Regional Government, Iraq. The authors appreciatively acknowledge the financial support from the Kurdistan National Research Council (KNRC)- Ministry of Higher Education and Scientific Research-KRG/Iraq for this research project.

Conflicts of Interest: The authors declare no conflict of interest.

References

1. Ambasankar, K.N.; Bhattacharjee, L.; Jat, S.K.; Bhattacharjee, R.R.; Mohanta, K. Study of Electrical Charge Storage in Polymer-Carbon Quantum Dot Composite. *Chem. Sel.* **2017**, *2*, 4241–4247. [CrossRef]
2. Wang, R.; Lu, K.-Q.; Tang, Z.-R.; Xu, Y.-J. Recent progress in carbon quantum dots: Synthesis, properties and applications in photocatalysis. *J. Mater. Chem. A* **2017**, *5*, 3717–3734. [CrossRef]
3. Atabaev, T.S. Doped Carbon Dots for Sensing and Bioimaging Applications: A Minireview. *Nanomaterials* **2018**, *8*, 342. [CrossRef] [PubMed]
4. Yang, G.; Wan, X.; Liu, Y.; Li, R.; Su, Y.; Zeng, X.; Tang, J. Luminescent poly(vinyl alcohol)/carbon quantum dots composites with tunable water-induced shape memory behavior in different pH and temperature environments. *ACS Appl. Mater. Interfaces* **2016**, *8*, 34744–34754. [CrossRef] [PubMed]
5. Choi, Y.; Jeon, D.; Choi, Y.; Ryu, J.; Kim, B.-S. Self-Assembled Supramolecular Hybrid of Carbon Nanodots and Polyoxometalates for Visible-Light-Driven Water Oxidation. *ACS Appl. Mater. Interfaces* **2018**, *10*, 13434–13441. [CrossRef] [PubMed]
6. Kovalchuk, A.; Huang, K.; Xiang, C.; Martí, A.A.; Tour, J.M. Luminescent Polymer Composite Films Containing Coal-Derived Graphene Quantum Dots. *ACS Appl. Mater. Interfaces* **2015**, *7*, 26063–26068. [CrossRef] [PubMed]
7. Woelfle, C.; Claus, R.O. Transparent and flexible quantum dot–polymer composites using an ionic liquid as compatible polymerization medium. *Nanotechnology* **2007**, *18*. [CrossRef]
8. Suo, B.; Su, X.; Wu, J.; Chen, D.; Wang, A.; Guo, Z. Poly (vinyl alcohol) thin film filled with CdSe–ZnS quantum dots: Fabrication, characterization and optical properties. *Mater. Chem. Phys.* **2010**, *119*, 237–242. [CrossRef]
9. Azmer, M.I.; Ahmad, Z.; Sulaiman, K.; Touati, F. Morphological and structural properties of VoPcPhO:P3HT composite thin films. *Mater. Lett.* **2016**, *164*, 605–608. [CrossRef]
10. Aziz, S.B.; Abdullah, R.M.; Rasheed, M.A.; Ahmed, H.M. Role of Ion Dissociation on DC Conductivity and Silver Nanoparticle Formation in PVA:AgNt Based Polymer Electrolytes: Deep Insights to Ion Transport Mechanism. *Polymers* **2017**, *9*, 338. [CrossRef]
11. Aziz, S.B.; Abdulwahid, R.T.; Rasheed, M.A.; Abdullah, O.G.; Ahmed, H.M. Polymer Blending as a Novel Approach for Tuning the SPR Peaks of Silver Nanoparticles. *Polymers* **2017**, *9*, 486. [CrossRef]
12. Chen, L.; Zhang, C.; Du, Z.; Li, H.; Zhang, L.; Zou, W. Fabrication of amido group functionalized carbon quantum dots and its transparent luminescent epoxy matrix composites. *J. Appl. Polym. Sci.* **2015**, *132*, 42667. [CrossRef]
13. Bhunia, S.K.; Nandi, S.; Shikler, R.; Jelinek, R. Tuneable light-emitting carbon-dot/polymer flexible films prepared through one-pot synthesis. *Nanoscale* **2016**, *8*, 3400–3406. [CrossRef] [PubMed]
14. Ahad, N.; Saion, E.; Gharibshahi, E. Structural, Thermal, and Electrical Properties of PVA-Sodium Salicylate Solid Composite Polymer Electrolyte. *J. Nanomater.* **2012**, *2012*, 857569. [CrossRef]
15. Bhargav, P.B.; Mohan, V.M.; Sharma, A.K.; Rao, V.V.R.N. Structural, Electrical and Optical Characterization of Pure and Doped Poly (Vinyl Alcohol) (PVA) Polymer Electrolyte Films. *Int. J. Polym. Mater.* **2007**, *56*, 579–591. [CrossRef]
16. Gao, H.; Lian, K. Characterizations of proton conducting polymer electrolytes for electrochemical capacitors. *Electrochim. Acta* **2010**, *56*, 122–127. [CrossRef]
17. Radha, K.P.; Selvasekarapandian, S.; Karthikeyan, S.; Hema, M.; Sanjeeviraja, C. Synthesis and impedance analysis of proton-conducting polymer electrolyte PVA:NH$_4$F. *Ionics* **2013**, *10*, 1437–1447. [CrossRef]

18. Aziz, S.B.; Rasheed, M.A.; Hussein, A.M.; Ahmed, H.M. Fabrication of polymer blend composites based on [PVA-PVP](1−x):(Ag2S)x (0.01 ≤ x ≤ 0.03) with small optical band gaps: Structural and optical properties. *Mater. Sci. Semicond. Process.* **2017**, *71*, 197–203. [CrossRef]

19. Jiang, L.; Yang, T.; Peng, L.; Dan, Y. Acrylamide modified poly(vinyl alcohol): Crystalline and enhanced water solubility. *RSC Adv.* **2015**, *5*, 86598–86605. [CrossRef]

20. Liew, C.-W.; Arifin, K.H.; Kawamura, J.; Iwai, Y.; Ramesh, S.; Arof, A.K. Electrical and structural studies of ionic liquid-based poly(vinyl alcohol)proton conductors. *J. Non-Cryst. Solids* **2015**, *425*, 163–172. [CrossRef]

21. Omer, K.M.; Hassan, A.Q. Chelation-enhanced fluorescence of phosphorus doped carbon nanodots for multi-ion detection. *Microchim. Acta* **2017**, *184*, 2063–2071. [CrossRef]

22. Aziz, S.B. Modifying poly (vinyl alcohol)(PVA) from insulator to small-bandgap polymer: A novel approach for organic solar cells and optoelectronic devices. *J. Electron. Mater.* **2016**, *45*, 736–745. [CrossRef]

23. Dong, Y.; Lin, J.G.; Chen, Y.; Fu, F.; Chi, Y.; Chen, G. Graphene quantum dots, graphene oxide, carbon quantum dots and graphite nanocrystals in coals. *Nanoscale* **2014**, *6*, 7410–7415. [CrossRef] [PubMed]

24. Saranya, M.; Ramachandranb, R.; Samuel, E.J.J.; Jeong, S.K.; Grace, A.N. Enhanced visible light photocatalytic reduction of organic pollutant and electrochemical properties of CuS catalyst. *Powder Technol.* **2015**, *279*, 209–220. [CrossRef]

25. Li, Y.; Scott, J.; Chen, Y.; Guo, L.; Zhao, M.; Wang, X.; Lu, W. Direct dry-grinding synthesis of monodisperse lipophilic CuSnanoparticles. *Mater. Chem. Phys.* **2015**, *162*, 671–676. [CrossRef] [PubMed]

26. Aziz, S.B.; Abdulwahid, R.T.; Rsaul, H.A.; Ahmed, H.M. In situ synthesis of CuS nanoparticle with a distinguishable SPR peak in NIR region. *J. Mater. Sci. Mater. Electron.* **2016**, *27*, 4163–4171. [CrossRef]

27. Aravindan, V.; Lakshmi, C.; Vickraman, P. Investigations on Na+ ion conducting polyvinylidenefluoride-co-hexafluoro-propylene/poly ethylmethacrylate blend polymer electrolytes. *Curr. Appl. Phys.* **2009**, *9*, 1106–1111. [CrossRef]

28. Aziz, S.B.; Abidin, Z.H.Z.; Kadir, M.F.Z. Innovative method to avoid the reduction of silver ions to silver nanoparticles (Ag+ → Ag°)in silver ion conducting based polymer electrolytes. *Phys. Scr.* **2015**, *90*, 035808. [CrossRef]

29. Aziz, S.B.; Abdullah, O.G.; Rasheed, M.A. Structural and electrical characteristics of PVA: NaTf based solid polymer electrolytes: Role of lattice energy of salts on electrical DC conductivity. *J. Mater. Sci. Mater. Electron.* **2017**, *28*, 12873–12884. [CrossRef]

30. Aziz, S.B.; Abdullah, R.M.; Kadir, M.F.Z.; Ahmed, H.M. Non suitability of silver ion conducting polymer electrolytes based on chitosan mediated by barium titanate (BaTiO3) for electrochemical device applications. *Electrochim. Acta* **2019**, *296*, 494–507. [CrossRef]

31. Aziz, S.B. Role of dielectric constant on ion transport: Reformulated Arrhenius equation. *Adv. Mater. Sci. Eng.* **2016**, *2016*, 2527013. [CrossRef]

32. Wei, D.; Sun, W.; Qian, W.; Ye, Y.; Ma, X. The synthesis of chitosan-based silver nanoparticles and their antibacterial activity. *Carbohydr. Res.* **2009**, *344*, 2375–2382. [CrossRef] [PubMed]

33. Hess, S.C.; Permatasari, F.A.; Fukazawa, H.; Schneider, E.M.; Balgis, R.; Ogi, T.; Okuyama, K.; Stark, W.J. Direct synthesis of carbon quantum dots in aqueous polymer solution: One-pot reaction and preparation of transparent UV-blocking films. *J. Mater. Chem. A* **2017**, *5*, 5187–5194. [CrossRef]

34. Kumar, R.; Ali, S.A.; Mahur, A.K.; Virk, H.S.; Singh, F.; Khan, S.A.; Avasthi, D.K.; Prasad, R. Study of optical band gap and carbonaceous clusters in swift heavy ion irradiated polymers with UV–Vis spectroscopy. *Nucl. Instrum. Methods Phys. Res. B* **2008**, *266*, 1788–1792. [CrossRef]

35. Emam, A.N.; Loutfy, S.A.; Mostafa, A.A.; Awad, H.; Mohamed, M.B. Cyto-toxicity, biocompatibility and cellular response of carbon dots–plasmonic based nano-hybrids for bioimaging. *RSC Adv.* **2017**, *7*, 23502–23514. [CrossRef]

36. Li, X.; Zhang, S.; Kulinich, S.A.; Liu, Y.; Zeng, H. Engineering surface states of carbon dots to achieve controllable luminescence for solid-luminescent composites and sensitive Be21 detection. *Sci. Rep.* **2014**, *4*, 4976. [CrossRef]

37. Wang, Y.; Hu, A. Carbon quantum dots: Synthesis, properties and applications. *J. Mater. Chem. C* **2014**, *2*, 6921–6939. [CrossRef]

38. Weaver, J.; Zakeri, R.; Aouadib, S.; Kohli, P. Synthesis and characterization of quantum dot–polymer composites. *J. Mater. Chem.* **2009**, *19*, 3198–3206. [CrossRef]

39. Stan, C.S.; Horlescu, P.G.; Ursu, L.E.; Popa, M.; Albu, C. Facile preparation of highly luminescent composites by polymer embedding of carbon dots derived from N-hydroxyphthalimide. *J. Mater. Sci.* **2017**, *52*, 185–196. [CrossRef]

40. Saq'an, S.A.; Ayesh, A.S.; Zihlif, A.M.; Martuscelli, E.; Ragosta, G. Physical properties of polystyrene/alum composites. *Polym. Test.* **2004**, *23*, 739–745. [CrossRef]

41. Aziz, S.B.; Abdullah, O.G.; Hussein, A.M.; Ahmed, H.M. From Insulating PMMA Polymer to Conjugated Double Bond Behavior: Green Chemistry as a Novel Approach to Fabricate Small Band Gap Polymers. *Polymers* **2017**, *9*, 626. [CrossRef]

42. Hareesh, K.; Sanjeev, G.; Pandey, A.K.; Rao, V. Characterization of UV-irradiated Lexan polycarbonate films. *Iran Polym. J.* **2013**, *22*, 341–349. [CrossRef]

43. Wang, H.; Ferrio, K.; Steel, D.; Hu, Y.; Binder, R.; Koch, S.W. Transient nonlinear optical response from excitation induced dephasing in GaAs. *Phys. Rev. Lett.* **1993**, *71*, 1261–1264. [CrossRef] [PubMed]

44. Aziz, S.B.; Ahmed, H.M.; Hussein, A.M.; Fathulla, A.B.; Wsw, R.M.; Hussein, R.T. Tuning the absorption of ultraviolet spectra and optical parameters of aluminum doped PVA based solid polymer composites. *J. Mater. Sci. Mater. Electron.* **2015**, *26*, 8022–8028. [CrossRef]

45. Aziz, S.B.; Abdullah, O.G.; Hussein, A.M.; Abdulwahid, R.T.; Rasheed, M.A.; Ahmed, H.M.; Abdalqadir, S.W.; Mohammed, A.R. Optical properties of pure and doped PVA: PEO based solid polymer blend electrolytes: Two methods for band gap study. *J. Mater. Sci. Mater. Electron.* **2017**, *28*, 7473–7479. [CrossRef]

46. Babu, K.E.; Veeraiah, A.; Swamy, D.T.; Veeraiah, V. First-principles study of electronic and optical properties of cubic perovskite CsSrF3. *Mater. Sci. Pol.* **2012**, *30*, 359–367. [CrossRef]

47. Yakuphanoglua, F.; Arslan, M. Determination of thermo-optic coefficient, refractive index, optical dispersion and group velocity parameters of an organic thin film. *Phys. B* **2007**, *393*, 304–309. [CrossRef]

48. Zhou, Y.; Sharma, S.K.; Peng, Z.; Leblanc, R.M. Polymers in Carbon Dots: A Review. *Polymers* **2017**, *9*, 67. [CrossRef]

49. Jin, J.; Qi, R.; Su, Y.; Tong, M.; Zhu, J. Preparation of high-refractive-index PMMA/TiO2 nanocomposites by one-step in situ solvothermal method. *Iran Polym. J.* **2013**, *22*, 767–774. [CrossRef]

50. Tao, P.; Li, Y.; Rungta, A.; Viswanath, A.; Gao, J.; Benicewicz, B.C.; Siegel, R.W.; Schadler, L.S. TiO$_2$ nanocomposites with high refractive index and transparency. *J. Mater. Chem.* **2011**, *21*, 18623–18629. [CrossRef]

51. Aziz, S.B.; Rasheed, M.A.; Ahmed, H.M. Synthesis of Polymer Nanocomposites Based on [Methyl Cellulose](1− x):(CuS) x (0.02 M≤ x≤ 0.08 M) with Desired Optical Band Gaps. *Polymers* **2017**, *9*, 194. [CrossRef]

52. Aziz, S.B. Morphological and Optical Characteristics of Chitosan (1−x):Cuox (4 ≤ x ≤ 12) Based Polymer Nano-Composites: Optical Dielectric Loss as an Alternative Method for Tauc's Model. *Nanomaterials* **2017**, *7*, 444. [CrossRef]

53. Saini, I.; Rozra, J.; Chandak, N.; Aggarwal, S.; Sharma, P.K.; Sharma, A. Tailoring of electrical, optical and structural properties of PVA by addition of Ag nanoparticles. *Mater. Chem. Phys.* **2013**, *139*, 802–810. [CrossRef]

54. Biskri, Z.E.; Rached, H.; Bouchear, M.; Rached, D.; Aida, M.S. A Comparative Study of Structural Stability and Mechanical and Optical Properties of Fluorapatite (Ca5(PO4)3F) and Lithium Disilicate (Li2Si2O5) Components Forming Dental Glass–Ceramics: First Principles Study. *J. Electron. Mater.* **2016**, *45*, 5082–5095. [CrossRef]

55. Ravindra, N.M.; Ganapathy, P.; Choi, J. Energy gap–refractive index relations in semiconductors—An overview. *Infrared Phys. Technol.* **2007**, *50*, 21–29. [CrossRef]

56. Plass, M.F.; Popov, C.; Ivanov, B.; Mänd, S.; Jelinek, M.; Zambov, L.M.; Kulisch, W. Correlation between photoluminescence, optical and structuralproperties of amorphous nitrogen-rich carbon nitride films. *Appl. Phys. A* **2001**, *72*, 21–27. [CrossRef]

57. Aziz, S.B.; Rasheed, M.A.; Abidin, Z.H.Z. Optical and electrical characteristics of silver ion conducting nanocomposite solid polymer electrolytes based on chitosan. *J. Electron. Mater.* **2017**, *46*, 6119. [CrossRef]

58. Aziz, S.B.; Mamand, S.M.; Saed, S.R.; Abdullah, R.M.; Hussein, S.A. New Method for the Development of Plasmonic Metal-Semiconductor Interface Layer: Polymer Composites with Reduced Energy Band Gap. *J. Nanomater.* **2017**, *2017*, 8140693. [CrossRef]

59. He, C.-E.; Zeng, Z.-Y.; Cheng, Y.; Chen, X.-R.; Cai, L.C. First-principles calculations for electronic, optical and thermodynamic properties of ZnS. *Chin. Phys. B* **2008**, *17*, 3867–3874.

60. Cheddadi, S.; Boubendira, K.; Meradji, H.; Ghemid, S.; Hassan, F.E.H.; Lakel, S.; Khenata, R. First-principle calculations of structural, electronic, optical, elastic and thermal properties of MgXAs2 (X = Si, Ge) compounds. *Pramana J. Phys.* **2017**, *89*, 89. [CrossRef]

61. Nasr, T.B.; Maghraoui-Meherzi, H.; Abdallah, H.B.; Bennaceur, R. First principles calculations of electronic and optical properties of Ag2S. *Solid State Sci.* **2013**, *26*, 65–71. [CrossRef]

62. Rahman, M.A.; Rahman, M.A.; Chowdhury, U.K.; Bhuiyan, M.T.H.; Ali, M.L.; Sarker, M.A.R. First principles investigation of structural, elastic, electronic and optical properties of ABi2O6 (A = Mg,Zn) with trirutile-type structure. *Cogent Phys.* **2016**, *3*, 1257414. [CrossRef]

63. Xin-Yin, Z.; Yue-Hua, W.; Min, Z.; Na, Z.; Sai, G.; Qiong, C. First-Principles Calculations of the Structural, Electronic and Optical Properties of BaZrxTi1−xO3 (x= 0, 0.25, 0.5, 0.75). *Chin. Phys. Lett.* **2011**, *28*, 067101.

64. Ching, W.Y.; Gu, Z.-Q.; Xu, Y.-N. First-principles calculation of the electronic and optical properties of LiNb03. *Phys. Rev. B* **1994**, *50*, 1992–1995. [CrossRef]

65. Aziz, S.B.; Abdullah, O.G.; Rasheed, M.A. A novel polymer composite with a small optical band gap: New approaches for photonics and optoelectronics. *J. Appl. Polym. Sci.* **2017**, *134*, 44847. [CrossRef]

66. Li, X.; Cui, H.; Zhang, R. First-principles study of the electronic and optical properties of a new metallic MoAlB. *Sci. Rep.* **2016**, *6*, 39790. [CrossRef]

67. Huang, J.J.; Zhong, Z.F.; Rong, M.Z.; Zhou, X.; Chen, X.D.; Zhang, M.Q. An easy approach of preparing strongly luminescent carbon dots and their polymer based composites for enhancing solar cell efficiency. *Carbon* **2014**, *70*, 190–198. [CrossRef]

68. Jiang, Z.C.; Lin, T.N.; Lin, H.T.; Talite, M.J.; Tzeng, T.T.; Hsu, C.L.; Chiu, K.P.; Lin, C.A.J.; Shen, J.L.; Yuan, C.T. A Facile and Low-Cost Method to Enhance the Internal Quantum Yield and External Light-Extraction Efficiency for Flexible Light-Emitting Carbon-Dot Films. *Sci. Rep.* **2016**, *6*, 19991. [CrossRef]

69. Zidan, H.M.; Abu-Elnader, M. Structural and optical properties of pure PMMA and metal chloride-doped PMMA films. *Phys. B* **2005**, *355*, 308–317. [CrossRef]

70. Huang, J.R.; Her, S.-C.; Yang, X.X.; Zhi, M.N. Synthesis and Characterization of Multi-Walled Carbon Nanotube/Graphene Nanoplatelet Hybrid Film for Flexible Strain Sensors. *Nanomaterials* **2018**, *8*, 786. [CrossRef]

71. Ke, C.-B.; Lu, T.-L.; Chen, J.-L. Capacitively Coupled Plasma Discharge of Ionic Liquid Solutions to Synthesize Carbon Dots as Fluorescent Sensors. *Nanomaterials* **2018**, *8*, 372. [CrossRef] [PubMed]

72. Genc, R.; Alas, M.O.; Harputlu, E.; Repp, S.; Kreme, N.; Castellano, M.; Colak, S.G.; Ocakoglu, K.; Erd, E. High-Capacitance Hybrid Supercapacitor Based on Multi-Colored Fluorescent Carbon-Dots. *Sci. Rep.* **2017**, *7*, 11222. [CrossRef] [PubMed]

Article

Exploring the Role of Nanoparticles in Enhancing Mechanical Properties of Hydrogel Nanocomposites

Josergio Zaragoza [1], Scott Fukuoka [1], Marcus Kraus [1], James Thomin [2] and Prashanth Asuri [1,*]

[1] Department of Bioengineering, Santa Clara University, Santa Clara, CA 95053, USA;
j1zaragoza@scu.edu (J.Z.); sfukuoka@scu.edu (S.F.); mkraus1@scu.edu (M.K.)

[2] Department of General Sciences, Northwest Florida State College, Niceville, FL 32578, USA;
thominj@nwfsc.edu

* Correspondence: asurip@scu.edu; Tel.: +1-408-551-3005

Received: 27 September 2018; Accepted: 25 October 2018; Published: 29 October 2018

Abstract: Over the past few decades, research studies have established that the mechanical properties of hydrogels can be largely impacted by the addition of nanoparticles. However, the exact mechanisms behind such enhancements are not yet fully understood. To further explore the role of nanoparticles on the enhanced mechanical properties of hydrogel nanocomposites, we used chemically crosslinked polyacrylamide hydrogels incorporating silica nanoparticles as the model system. Rheological measurements indicate that nanoparticle-mediated increases in hydrogel elastic modulus can exceed the maximum modulus that can be obtained through purely chemical crosslinking. Moreover, the data reveal that nanoparticle, monomer, and chemical crosslinker concentrations can all play an important role on the nanoparticle mediated-enhancements in mechanical properties. These results also demonstrate a strong role for pseudo crosslinking facilitated by polymer–particle interactions on the observed enhancements in elastic moduli. Taken together, our work delves into the role of nanoparticles on enhancing hydrogel properties, which is vital to the development of hydrogel nanocomposites with a wide range of specific mechanical properties.

Keywords: hydrogel nanocomposites; elastic modulus; rotational rheology; pseudo-crosslinking

---

1. Introduction

Hydrogels have recently emerged as potential candidates for various biomedical and biotechnological applications owing to their unique physical and biochemical properties. Composed of highly porous and hydrated networks, they allow cell encapsulation for tissue engineering applications and support the loading and release of various bioactive molecules for drug delivery applications [1–3]. The use of hydrogels in bioseparations and biosensing and as tissue-adhesives has also been recently proposed [4,5]. Latest advances in polymer chemistry and synthesis as well as progress in the development of interpenetrating polymer network hydrogels have opened up new possibilities in developing hydrogel-based biomaterials with advanced properties. However, their poor mechanical properties have been a significant barrier to their widespread adoption for these applications. Over three decades of research have shown that the addition of nanoscopic filler particles to a variety of polymer systems (melts, elastomers, hydrogels, etc.) can have a large effect on their mechanical properties [6–9]. Both experimental studies [10–12] and modeling analyses [13–15] have indicated that the enhancements in nanocomposite properties relative to those of pure polymers are due in large part to an increase in polymer crosslink/entanglement density mediated by strong interactions with nanoparticles.

In this study, we investigate whether nanoparticle-mediated enhancements in polymer mechanical properties extend to hydrogel nanocomposites. Current research investigating the effects of incorporating nanoparticles reports enhanced mechanical properties for hydrogel nanocomposites relative to neat hydrogels [16–19]. However, barring a few studies [16,20], little research has been performed to understand the role of chemical or covalent crosslinking on nanoparticle-mediated changes in hydrogel properties. Such studies are crucial to developing a quantitative understanding of how pseudo-crosslinking or crosslinking density plays a role in the reinforcements observed due to the addition of nanoparticles. In our study, we performed rotational rheological measurements to evaluate the changes in the elastic modulus of chemically crosslinked polyacrylamide (pAAm) hydrogels due to the addition of silica nanoparticles. pAAm hydrogels are synthesized as a networked structure of repeating acrylamide (AAm) subunits chemically crosslinked using the bifunctional crosslinking agent N,N'-methylenebisacrylamide (Bis). Silica nanoparticles (SiNPs) are known to hydrogen bond with polyacrylamide [21,22] and so we expect there to be a strong interfacial binding energy between the polymer and nanoparticle surface. Strong interactions between nanoparticles and polymer chains have been shown to facilitate nanoparticle-mediated reinforcement of hydrogels in previous studies [23]. Therefore, this system allowed us to evaluate the effects of nanoparticle-mediated physical crosslinking in comparison to chemical crosslinking, by varying the degree of both chemical- and nanoparticle-mediated crosslinking as well as the monomer concentration.

2. Results

2.1. Influence of Crosslinker Concentration on Hydrogel Elastic Modulus

Our initial experiments focused on the impact of crosslinker concentration on hydrogel modulus. pAAm hydrogels prepared using various monomer and crosslinker (AAm/Bis) ratios, as shown in Table 1, were characterized using rotational rheometry. When elastic modulus is plotted against the concentration of Bis, the hydrogel elastic modulus initially increases with crosslinker concentration for all concentrations of the monomer. However, above a threshold crosslinker concentration, the elastic modulus for each hydrogel reaches a plateau value (Table 1).

Table 1. Elastic modulus, G', for neat hydrogels for various monomer and crosslinker ratios.

%Bis	Elastic Modulus, G' [a]		
	10% AAm	5% AAm	2.5% AAm
1	$(1.96 \pm 0.14) \times 10^4$	$(3.72 \pm 0.13) \times 10^3$	$(4.22 \pm 0.42) \times 10^2$
0.5	$(1.98 \pm 0.36) \times 10^4$	$(3.57 \pm 0.19) \times 10^3$	$(4.07 \pm 0.49) \times 10^2$
0.25	$(1.03 \pm 0.03) \times 10^4$	$(3.67 \pm 0.27) \times 10^3$	$(4.27 \pm 0.22) \times 10^2$
0.125	$(5.78 \pm 0.86) \times 10^3$	$(1.95 \pm 0.09) \times 10^3$	$(4.14 \pm 0.59) \times 10^2$
0.0625	$(2.67 \pm 0.35) \times 10^3$	$(5.98 \pm 0.17) \times 10^2$	$(4.78 \pm 0.81) \times 10^1$

[a] Each data point represents an average of triplicate measurements.

Next, in Table 2, we show elastic modulus as a function of $\%C_{Bis}$, relative concentration of the crosslinker Bis, which is defined by Equation (1):

$$\%C_{Bis} = \frac{m_{Bis}}{m_{Bis} + m_{AAm}} \tag{1}$$

where m_{AAm} is the concentration of the monomer and m_{Bis} is the concentration of the crosslinker.

For each hydrogel, irrespective of the monomer concentration, the threshold point occurs at the same relative crosslinker concentration ($\%C_{Bis}$ = 4.76) (Supplementary Figure S1). Such saturation of hydrogel elastic modulus at a threshold relative chemical crosslinker concentration has been demonstrated previously [24], and has been shown to be the result of the formation of highly crosslinked "microgels" connected by a percolating network of linear polymer chains. Each microgel forms a mesoscale crosslink point, so that the overall crosslink density becomes a function of the number of microgels rather than the number of chemical crosslinking molecules. Since the kinetics of microgel formation are governed by the local concentration of the crosslinker, the number of microgels per unit volume reaches a maximum at a particular crosslinker concentration relative to the monomer [24,25].

Table 2. Elastic modulus, G′, for neat hydrogels for various monomer and relative crosslinker ratios.

%C_{Bis}	Elastic Modulus, G′ [a]		
	10% AAm	5% AAm	2.5% AAm
9.09	$(1.96 \pm 0.14) \times 10^4$	$(3.57 \pm 0.19) \times 10^3$	$(4.27 \pm 0.22) \times 10^2$
4.76	$(1.98 \pm 0.36) \times 10^4$	$(3.67 \pm 0.27) \times 10^3$	$(4.14 \pm 0.59) \times 10^2$
2.44	$(1.03 \pm 0.03) \times 10^4$	$(1.95 \pm 0.09) \times 10^3$	$(4.78 \pm 0.81) \times 10^1$
1.23	$(5.78 \pm 0.86) \times 10^3$	$(5.98 \pm 0.17) \times 10^2$	N/A [b]

[a] Each data point represents an average of triplicate measurements. [b] This condition was below the limits of reliable detection using rotational rheology.

2.2. Influence of Nanoparticles on the Elastic Modulus of Chemically Crosslinked Hydrogels

We proceeded to study the influence of nanoparticles on the chemically crosslinked pAAm hydrogels. When silica nanoparticles were added to the pAAm hydrogel, we observed a concentration-dependent increase in the elastic modulus (Figure 1a). Moreover, the experiments revealed that addition of nanoparticles increase the elastic modulus beyond the maximum modulus observed for the purely chemically crosslinked system (Figure 1b). It is interesting to note that even at sub-saturation values of the relative crosslinker concentration, incorporation of nanoparticles allows enhancements beyond the plateau modulus achieved using chemical crosslinking (Figure 1b). This behavior was observed for hydrogels over a range of monomer concentrations. These results are consistent with previous studies, including our investigations that have indicated hydrogen bonding mediated interactions between pAAm chains and SiNP surfaces enable silica nanoparticles to serve as pseudo crosslinkers and increase the extent of crosslinking in the hydrogel network, thereby facilitating reinforcements in the mechanical properties [22,26].

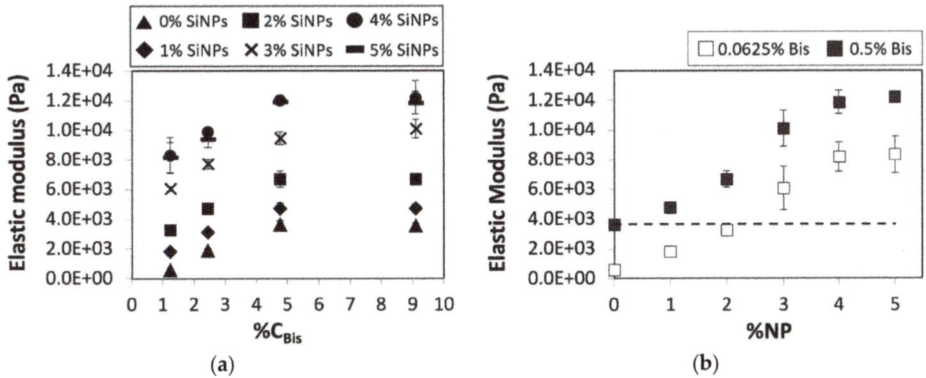

Figure 1. Elastic moduli of 5% pAAm hydrogels as a function of (**a**) relative crosslinker concentration (%C_{Bis}) prepared using different concentrations of 4 nm silica nanoparticles—0% SiNPs (triangles), 1% SiNPs (diamonds), 2% SiNPs (squares), 3% SiNPs (Xs), 4% SiNPs (dashes), and 5% SiNPs (circles) and (**b**) nanoparticle concentration (%NP) prepared using different concentrations of the chemical crosslinker—0.0625% Bis (open squares) and 0.5% Bis (closed squares). Data shown are the mean of triplicate measurements ± standard deviation and have been repeated at least three times with similar results. Dashed line in panel (**b**) represents elastic modulus of 5% pAAm hydrogel prepared using 0.5% Bis and 0% SiNPs and acts as a guide to the eye.

We also compared the effects of chemical crosslinking and nanoparticles on the viscous modulus, G″. These experiments revealed similar increases in the values of G′ and G″, when plotted against the concentration of Bis. However, we observed significant increases for G″ relative to G′, when plotted against the nanoparticle concentration (Figure 2a). Not surprisingly, the ratio of G″/G′ (tan δ) exhibited a significantly higher slope for tan δ plotted against the nanoparticle concentration compared to tan δ plotted against the concentration of Bis (Figure 2b). These experiments, therefore, indicate that nanoparticle-mediated reinforcements are independent from those mediated by chemical crosslinking.

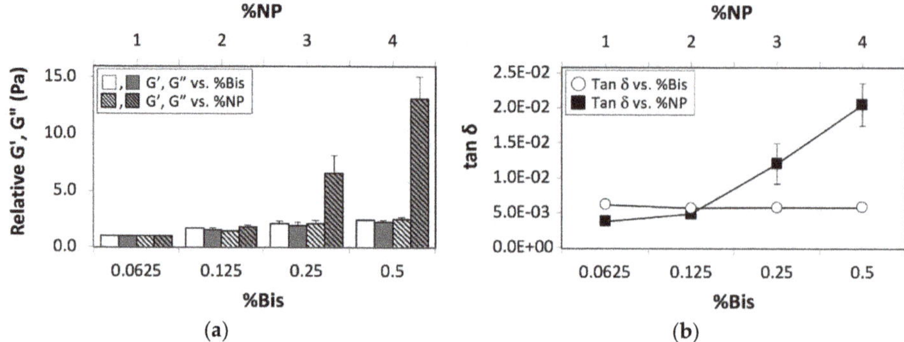

Figure 2. (**a**) Relative elastic (white bars) and viscous (grey bars) moduli of 5% pAAm hydrogels as a function of crosslinker concentration (%Bis) (solid bars) prepared using 1% 4 nm SiNPs and nanoparticle concentration (%NP) prepared using 0.0625% Bis (hashed bars), and (**b**) tan δ (ratio of G″/G′) as a function of the crosslinker (white circles) and nanoparticle concentration (black squares). Data shown are the mean of triplicate measurements ± standard deviation and have been repeated at least three times with similar results.

2.3. Influence of Chemical Crosslinking on Nanoparticle Mediated Enhancements of Hydrogel Elastic Modulus

Finally, we explored the combined effects of chemical and nanoparticle-mediated crosslinking on the pAAm elastic modulus. We compared the enhancements afforded by the incorporation of nanoparticles (G′$_{NP}$/G′$_0$, the ratio of elastic modulus of hydrogels that incorporated or did not incorporate nanoparticles) over a range of both crosslinker and monomer concentrations. Interestingly, these experiments revealed a diminishing impact of nanoparticles on the elastic modulus at higher chemical crosslinker and monomer concentrations (Figure 3 and Supplementary Figure S2). The decreasing role of nanoparticles with increasing crosslinker concentrations indicates a saturation point in the overall hydrogel crosslinking density and thereby enhancements in mechanical properties afforded by either covalent- or pseudo crosslinking and a combination thereof.

To better understand the role of monomer concentration on the nanoparticle-mediated enhancements in elastic modulus, we defined a new variable (%C_{NP}) that refers to the concentration of SiNPs relative to the monomer concentration in a manner similar to %C_{Bis} (Equation (2)).

$$\%C_{NP} = \frac{m_{NP}}{m_{NP} + m_{AAm}} \tag{2}$$

where m_{AAm} is the concentration of the monomer and m_{NP} is the concentration of the nanoparticles.

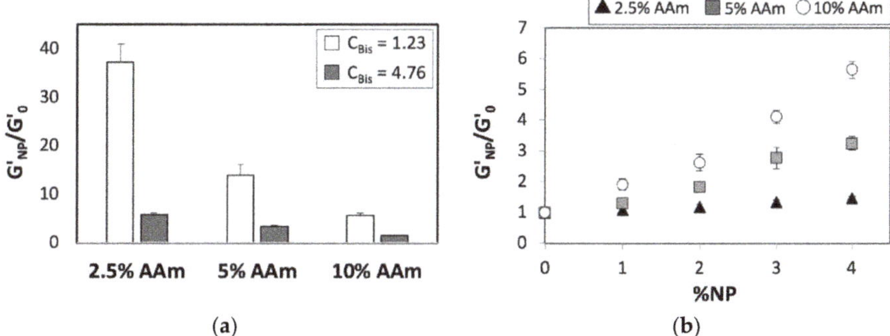

(**a**)   (**b**)

Figure 3. Relative elastic moduli of pAAm hydrogels as a function of (**a**) monomer concentration (%AAm) prepared using 5% 4 nm SiNPs and different concentrations of the chemical crosslinker—0.0625% Bis (white bars) and 0.5% Bis (grey bars) and (**b**) nanoparticle concentration for 2.5% pAAm (white circles), 5% pAAm (grey squares), and 10% pAAm (black triangles) hydrogels prepared using %C_{Bis} = 4.76. Data shown are the mean of triplicate measurements ± standard deviation and have been repeated at least three times with similar results.

The enhancements in elastic modulus afforded by the incorporation of nanoparticles for various AAm concentrations at saturation (%C_{Bis} = 4.76) and sub-saturation (%C_{Bis} = 1.23) concentrations of Bis, previously plotted against the absolute nanoparticle concentration (%NPs), were plotted against the relative nanoparticle concentration (%C_{NP}) (Figure 4 and Supplementary Figure S3). The enhancements mediated by nanoparticles for various monomer concentrations (Figure 3b) can be collapsed onto a single curve (Figure 4) upon the introduction of %C_{NP}, which accounts for relative reinforcement due to nanoparticles at specific concentrations of the chemical crosslinker. These results thereby indicate that the nanoparticle mediated enhancements in hydrogel modulus scales with the relative nanoparticle concentration, not unlike the enhancements mediated by chemical crosslinking.

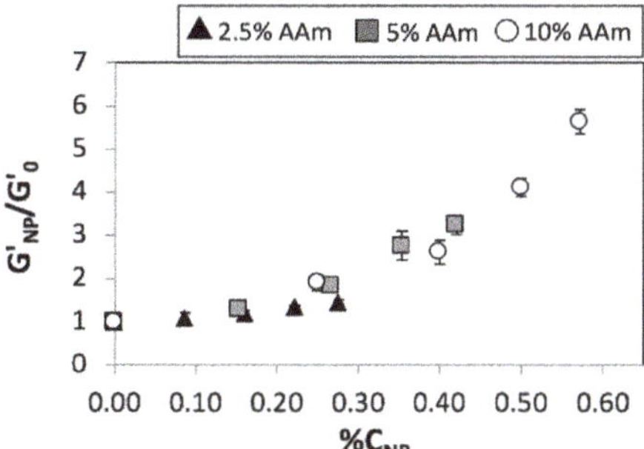

Figure 4. Relative elastic moduli of 2.5% pAAm (white circles), 5% pAAm (grey squares), and 10% pAAm (black triangles) hydrogels prepared using %C_{Bis} = 4.76 as a function of relative nanoparticle concentration (%C_{NP}). Data shown are the mean of triplicate measurements ± standard deviation and have been repeated at least three times with similar results.

2.4. Influence of Hydrophobic Side Chains on the Hydrogel Backbone on Nanoparticle Mediated Enhancements

To validate our hypothesis that suggests a saturation point in the overall hydrogel crosslinking density mediated by a combination of chemical (Bis) and physical (SiNPs) crosslinking, we repeated the mechanical characterization studies using poly(*N*-isopropylacrylamide) (pNIPAAm) hydrogels. pNIPAAm is a temperature-responsive hydrogel with a lower critical solution temperature (LCST) of ca. 32 °C in aqueous solutions [27,28]. The presence of hydrophobic side groups on the pNIPAAm hydrogel backbone leads to increased polymer–polymer interactions (and therefore increased physical crosslinking) at temperatures above LCST and thereby higher elastic modulus. Our initial experiments indicated that pNIPAAm hydrogels incorporating nanoparticles continue to exhibit temperature-dependent changes in mechanical properties, i.e., a reversible phase transition at temperatures greater than 32 °C. These experiments also indicated that pNIPAAm-SiNP hydrogels exhibit a higher elastic modulus compared to neat hydrogels (without nanoparticles). Interestingly, the nanoparticle-mediated enhancements in the elastic modulus were higher at temperatures <LCST relative to those at temperatures >LCST (Figure 5). This result is consistent with our hypothesis; if the reduced role of nanoparticles at higher crosslinker concentrations is a result of saturation in the overall hydrogel crosslinking density, then a similar effect (i.e., attenuated role of nanoparticles) should be observed in the presence of other sources of crosslinking as well. In the case of pNIPAAm hydrogels, an increase in physical crosslinking mediated by the hydrophobic side chains leads to saturation in overall crosslinking density and thereby an upper limit to enhancements achieved through the incorporation of nanoparticles.

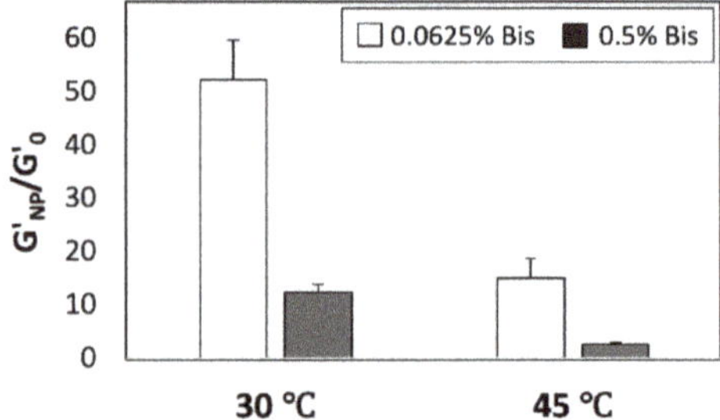

Figure 5. Relative elastic moduli of 5% pNIPAAm hydrogels prepared using 0.0625% Bis (white bars) or 0.5% Bis (grey bars), and 5% SiNPs at different temperatures. Data shown are the mean of triplicate measurements plus standard deviation and have been repeated at least three times with similar results.

3. Discussion

This study sought to examine the contributions of nanoparticle-mediated physical crosslinking to the elastic modulus of chemically crosslinked hydrogels, using pAAm-SiNP composites as the model system. Results from the rheological measurements showed that hydrogel elastic modulus positively correlated with both Bis and SiNP concentration. These experiments also indicated that a combination of chemical and physical crosslinking led to enhancements in elastic modulus that were greater than with either alone, which is consistent with previous studies [16,26,29]. Additional experiments suggested that the nanoparticle-mediated enhancements behave differently to those mediated by chemical crosslinking. However, we observed an upper limit to the gains in elastic modulus achievable through a combination of Bis- and SiNP-mediated crosslinking, suggesting the existence of a saturation point for the combined crosslinking density (i.e., the sum of crosslinking densities achieved through either chemical and physical means). To confirm the existence of a 'global' saturation point, beyond the 'local' saturation behaviors observed for either Bis- or SiNP-mediated enhancements, we introduced an additional source of physical crosslinking—pAAm functionalized with hydrophobic side groups that lead to increased polymer–polymer interactions at elevated temperatures. These experiments revealed attenuated enhancements mediated by nanoparticles at temperatures >LCST, consistent with the hypothesis that saturation in the overall crosslinking density can lead to upper limits in enhancements in modulus achieved through the addition of nanoparticles.

To better understand the observed role of nanoparticles on the hydrogel elastic modulus, we borrowed lessons learned from more classic polymer nanocomposite systems [6–8,10–15]. Certain traditional reinforcement models, such as simple filler effects, jamming theory, and fractal percolating networks, fail to explain the results from our studies. While we observe a linear relationship between elastic modulus and SiNP concentration for concentrations less than 3%, we observe saturation behavior at higher concentrations irrespective of monomer concentration. Furthermore, we have previously shown that the nanoparticle size also has an effect on the degree of enhancement, which means the enhancements cannot be explained by simple volume fraction arguments [30]. Jamming theory describes the response of a polymer to the drag friction caused by particles; however, it is unlikely for jamming to contribute significantly towards the elastic modulus of pAAm nanocomposites in our system given the dilute concentration of nanoparticles (0–5% SiNPs) [31,32]. The same can be said for fractal percolating structures that may form due to direct interactions among silica nanoparticles, as the likelihood that these arrangements will play a meaningful role at/below particle concentrations of 5% is low [33–35]. Other models that either explain reinforcement due to retardation of polymer dynamics in the interfacial zone around the nanoparticles, or the number of crosslinks or entanglements in a polymer network lend more support to our observations. In the cases where the polymer–particle interaction energy is much larger than the polymer–polymer interaction energy, the time scale of the movements of short lengths of polymer chains (the Rouse regime) is slowed down within a region extending from the particle surface to the limit of the polymer–particle energy well [13]. The result is an increase in elastic modulus that is proportional to the volume fraction of the interfacial zone and the magnitude of interfacial energy. Since volume fraction of the interfacial zone is positively correlated to particle concentration and negatively correlated to particle size, our results (i.e., increasing modulus with increasing SiNP concentration and decreasing SiNP size) can be explained by, albeit qualitatively, using the retardation polymer dynamics in the interfacial zone model. From a polymer–particle interaction standpoint, Flory's rubber elasticity theory may also provide a possible mechanism. Flory's network theory states that elastic modulus is proportional to the density of effective junctions [24,36]. As indicated in previous studies, non-covalent interactions between silica nanoparticles and polyacrylamide can serve to increase the number of effective crosslinks in the network and thereby lead to increases in elastic modulus, consistent with Flory's theory.

In addition to providing further insight into how nanoparticles mediate reinforcements in hydrogel elastic modulus, this study also contributes to the development of applications that may directly benefit from these improvements; examples include tissue engineering, drug delivery, wound dressings, and biosensing [17,37]. Some of these applications, especially those related to sensing, may also benefit from enhancements in the chemical and biological properties of the hydrogel. Nanoparticles may be functionalized with biomolecules prior to their incorporation into the hydrogel network to endow the hydrogels with specific biochemical characteristics in addition to improving their mechanical strength. Furthermore, new applications may also be realized by introducing and/or tailoring additional hydrogel properties such as thermal, electrical, optical, and magnetic characteristics. Several studies have already demonstrated the use of nanoparticles to develop polymer composites with conductive properties for battery cathodes and microelectronics [38], with light responsive properties for therapy [39], and with magnetic properties for electromagnetic interference shielding [40]. As we continue to further explore the use of nanoparticles to improve hydrogel properties, we can expect that the applications realized for traditional polymer composites may also translate to nanocomposites prepared using hydrogels.

4. Materials and Methods

4.1. Materials

All the materials for the polymerization reaction, acrylamide (AAm, monomer), initiator, ammonium persulfate (APS, initiator), *N*,*N*,*N'*,*N'*-tetramethylethylenediamine (TEMED, catalyst), and *N*,*N'*-methylenebis(acrylamide) (Bis, crosslinker), as well as *N*-isopropylacrylamide (NIPAAm, monomer for preparing thermoresponsive hydrogels) were purchased from Sigma Aldrich (St. Louis, MO, USA) and used as received. Tris-HCl buffer (pH 7.2) was obtained from Life Technologies

(Carlsbad, CA, USA) and binzil silica nanoparticle colloid solution with mean particle size of 4 nm was obtained as a gift from AkzoNobel Pulp and Performance Chemicals Inc. (Marietta, GA, USA).

4.2. Polymerization Reaction

Chemically crosslinked pAAm or pNIPAAm hydrogels were prepared as previously reported [26,41]. Briefly, the monomer (AAm or NIPAAm) and crosslinker (Bis) stocks were diluted to their desired concentrations in pH 7.2, 250 mM Tris-HCl buffer, followed by the addition of TEMED (0.1% of the final reaction volume) and 10% w/v APS solution (1% of the final reaction volume). For nanocomposite hydrogels, various amounts of silica nanoparticles (SiNPs) were added to the reaction mixture prior to the addition of APS and TEMED. Due to solubility limits, the maximum nanoparticle concentration used was 5% w/v. Polymerization reactions were performed at 25 °C between parallel plates of the rheometer cell to minimize exposure to air as oxygen inhibits the free radical polymerization reaction.

4.3. Measurement of Hydrogel Elastic Modulus

Rheological measurements of the hydrogels were carried out, as previously described, using the MCR302 rotational rheometer (Anton Paar, Graz, Austria) [26,41]. Briefly, 500 μL of a well-mixed reaction mixture was pipetted onto the lower plate of the rheometer and the upper plate was lowered until the desired gap distance (1 mm) was achieved. Amplitude sweeps at a constant frequency of 1 Hz were then carried out to ensure measurements were carried out in the linear viscoelastic regime of the hydrogels. Next, dynamic sweep tests over frequencies ranging from 0.1–100 Hz were recorded in the linear viscoelastic regimes (strain amplitude = 0.01) to determine the shear storage modulus. Final hydrogel parameters were determined by following the gelation for 90 min at 1 Hz and 1% strain for all samples. For the temperature studies using pNIPAAm hydrogels, the elastic modulus was measured at 30 °C and 45 °C using the conditions described above. Relative elastic and viscous moduli were calculated by normalizing the values for pAAm-SiNP hydrogels (G'_{NP} and G''_{NP}) to the corresponding values for control pAAm gels (G'_0 and G''_0).

Supplementary Materials: The following are available online http://www.mdpi.com/2079-4991/8/11/882/s1. Figure S1: Elastic modulus for neat hydrogels for various monomer and relative crosslinker ratios. Data shown are the mean of triplicate measurements ± standard deviation and have been repeated at least three times with similar results; Figure S2: Relative elastic moduli of pAAm hydrogels as a function of nanoparticle concentration for 2.5% pAAm (white circles), 5% pAAm (grey squares), and 10% pAAm (black triangles) hydrogels prepared using %C_{Bis} = 1.23. Data shown are the mean of triplicate measurements ± standard deviation and have been repeated at least three times with similar results; Figure S3: Relative elastic moduli of 2.5% pAAm (white circles), 5% pAAm (grey squares), and 10% pAAm (black triangles) hydrogels prepared using %C_{Bis} = 1.23 as a function of relative nanoparticle concentration (%CNP). Data shown are the mean of triplicate measurements ± standard deviation and have been repeated at least three times with similar results.

Author Contributions: P.A. conceived and designed the experiments; J.Z. and M.K. performed the experiments; J.T. and P.A. analyzed the data; S.F., J.T. and P.A. wrote the paper.

Funding: This research received no external funding.

Acknowledgments: This work was supported by the School of Engineering at Santa Clara University. Silica nanoparticles were a kind gift from AkzoNobel Pulp and Performance Chemicals Inc.

Conflicts of Interest: The authors declare no conflict of interest. The funding sponsors had no role in the design of the study, in the collection, analyses, or interpretation of data, in the writing of the manuscript, and in the decision to publish the results.

References

1. Liaw, C.Y.; Ji, S.; Guvendiren, M. Engineering 3D hydrogels for personalized in vitro human tissue models. *Adv. Healthc. Mater.* **2018**, *7*, 1–16. [CrossRef] [PubMed]
2. Peppas, N.A.; Huang, Y.; Torres-Lugo, M.; Ward, J.H.; Zhang, J. Physicochemical foundations and structural design of hydrogels in medicine and biology. *Annu. Rev. Biomed. Eng.* **2000**, *2*, 9–29. [CrossRef] [PubMed]

3. Garg, T.; Goyal, A.K. Biomaterial-based scaffolds–current status and future directions. *Expert Opin. Drug Deliv.* **2014**, *11*, 767–789. [CrossRef] [PubMed]

4. Jeong, B.; Gutowska, A. Lessons from nature: Stimuli-responsive polymers and their biomedical applications. *Trends Biotechnol.* **2002**, *20*, 305–311. [CrossRef]

5. Ghobril, C.; Grinstaff, M.W. The chemistry and engineering of polymeric hydrogel adhesives for wound closure: A tutorial. *Chem. Soc. Rev.* **2015**, *44*, 1820–1835. [CrossRef] [PubMed]

6. Jordan, J.; Jacob, K.I.; Tannenbaum, R.; Sharaf, M.A.; Jasiuk, I. Experimental trends in polymer nanocomposites—A review. *Mater. Sci. Eng. A* **2005**, *393*, 1–11. [CrossRef]

7. Tjong, S.C. Structural and mechanical properties of polymer nanocomposites. *Mater. Sci. Eng. R Rep.* **2006**, *53*, 73–197. [CrossRef]

8. Münstedt, H.; Triebel, C. Elastic properties of polymer melts filled with nanoparticles. *AIP Conf. Proc.* **2011**, *1375*, 201–207. [CrossRef]

9. Blanco, I. The Rediscovery of POSS: A Molecule Rather than a Filler. *Polymers* **2018**, *10*, 904. [CrossRef]

10. Georgopanos, P.; Schneider, G.A.; Dreyer, A.; Handge, U.A.; Filiz, V.; Feld, A.; Feld, A.; Yilmaz, E.D.; Krekeler, T.; Ritter, M.; et al. Exceptionally strong, stiff and hard hybrid material based on an elastomer and isotropically shaped ceramic nanoparticles. *Sci. Rep.* **2017**, *7*, 1–9. [CrossRef] [PubMed]

11. Mangal, R.; Srivastava, S.; Archer, L.A. Phase stability and dynamics of entangled polymer-nanoparticle composites. *Nat. Commun.* **2015**, *6*, 1–9. [CrossRef] [PubMed]

12. Langley, N.R.; Polmanteer, K.E. Relation of elastic modulus to crosslink and entanglement concentrations in rubber networks. *J. Polym. Sci. Polym. Phys. Ed.* **1974**, *12*, 1023–1034. [CrossRef]

13. Zeng, Q.H.; Yu, A.B.; Lu, G.Q. Multiscale modeling and simulation of polymer nanocomposites. *Prog. Polym. Sci.* **2008**, *33*, 191–269. [CrossRef]

14. Sen, S.; Thomin, J.D.; Kumar, S.K.; Keblinski, P. Molecular underpinnings of the mechanical reinforcement in polymer nanocomposites. *Macromolecules.* **2007**, *40*, 4059–4067. [CrossRef]

15. Frankland, S.J.V.; Caglar, A.; Brenner, D.W.; Griebel, M. Molecular simulation of the influence of chemical cross-links on the shear strength of carbon nanotube-polymer interfaces. *J. Phys. Chem. B* **2002**, *106*, 3046–3048. [CrossRef]

16. Adibnia, V.; Taghavi, S.M.; Hill, R.J. Roles of chemical and physical crosslinking on the rheological properties of silica-doped polyacrylamide hydrogels. *Rheol. Acta* **2017**, *56*, 123–134. [CrossRef]

17. Gaharwar, A.K.; Peppas, N.A.; Khademhosseini, A. Nanocomposite hydrogels for biomedical applications. *Biotechnol. Bioeng.* **2014**, *111*, 441–453. [CrossRef] [PubMed]

18. Yang, J.; Han, C.R.; Duan, J.F.; Xu, F.; Sun, R.C. Mechanical and viscoelastic properties of cellulose nanocrystals reinforced poly(ethylene glycol) nanocomposite hydrogels. *ACS Appl. Mater. Interfaces* **2013**, *5*, 3199–3207. [CrossRef] [PubMed]

19. Gaharwar, A.K.; Dammu, S.A.; Canter, J.M.; Wu, C.J.; Schmidt, G. Highly extensible, tough, and elastomeric nanocomposite hydrogels from poly(ethylene glycol) and hydroxyapatite nanoparticles. *Biomacromolecules* **2011**, *12*, 1641–1650. [CrossRef] [PubMed]

20. Lam, J.; Kim, K.; Lu, S.; Tabata, Y.; Scott, D.W.; Mikos, A.G.; Kasper, F.K. A factorial analysis of the combined effects of hydrogel fabrication parameters on the in vitro swelling and degradation of oligo(poly(ethylene glycol) fumarate) hydrogels. *J. Biomed. Mater. Res. Part A.* **2014**, *102*, 3477–3487. [CrossRef] [PubMed]

21. Lu, X.; Mi, Y. Characterization of the interfacial interaction between polyacrylamide and silicon substrate by fourier transform infrared spectroscopy. *Macromolecules* **2005**, *38*, 839–843. [CrossRef]

22. Wu, L.; Zeng, L.; Chen, H.; Zhang, C. Effects of silica sol content on the properties of poly(acrylamide)/silica composite hydrogel. *Polym. Bull.* **2012**, *68*, 309–316. [CrossRef]

23. Hu, X.; Wang, T.; Xiong, L.; Wang, C.; Liu, X.; Tong, Z. Preferential adsorption of poly(ethylene glycol) on hectorite clay and effects on poly(*N*-isopropylacrylamide)/hectorite nanocomposite hydrogels. *Langmuir* **2010**, *26*, 4233–4238. [CrossRef] [PubMed]

24. Calvet, D.; Wong, J.Y.; Giasson, S. Rheological monitoring of polyacrylamide gelation: Importance of cross-link density and temperature. *Macromolecules* **2004**, *37*, 7762–7771. [CrossRef]

25. Baselga, J.; Llorente, M.A.; Hernández-Fuentes, I.; Piérola, I.F. Polyacrylamide gels. Process of network formation. *Eur. Polym. J.* **1989**, *25*, 477–480. [CrossRef]

26. Zaragoza, J.; Babhadiashar, N.; O'Brien, V.; Chang, A.; Blanco, M.; Zabalegui, A.; Lee, H.; Asuri, P. Experimental investigation of mechanical and thermal properties of silica nanoparticle-reinforced poly(acrylamide) nanocomposite hydrogels. *PLoS ONE* **2015**, *10*, 1–11. [CrossRef] [PubMed]

27. Schild, H.G. Poly(*N*-isopropylacrylamide): Experiment, theory and application. *Prog. Polym. Sci.* **1992**, *17*, 163–249. [CrossRef]

28. Heskins, M.; Guillet, J.E. Solution Properties of Poly(*N*-isopropylacrylamide). *J. Macromol. Sci. Part A Chem.* **1968**, *2*, 1441–1455. [CrossRef]

29. Wu, C.J.; Gaharwar, A.K.; Chan, B.K.; Schmidt, G. Mechanically tough Pluronic F127/Laponite nanocomposite hydrogels from covalently and physically cross-linked networks. *Macromolecules.* **2011**, *44*, 8215–8224. [CrossRef]

30. Ulrich, Z. A simple theory of filler reinforcement in elastomers subjected to shear. *J. Appl. Polym. Sci.* **1966**, *10*, 1315–1322. [CrossRef]

31. Richter, S.; Kreyenschulte, H.; Saphiannikova, M.; Götze, T.; Heinrich, G. Studies of the so-called jamming phenomenon in filled rubbers using dynamical-mechanical experiments. *Macromol. Symp.* **2011**, *306–307*, 141–149. [CrossRef]

32. Trappe, V.; Prasad, V.; Cipelletti, L.; Segre, P.N.; Weitz, D.A. Jamming phase diagram for attractive particles. *Nature* **2001**, *411*, 772–775. [CrossRef] [PubMed]

33. Baeza, G.P.; Dessi, C.; Costanzo, S.; Zhao, D.; Gong, S.; Alegria, A.; Colby, R.H.; Rubinstein, M.; Vlassopoulos, D.; Kumar, S.K. Network dynamics in nanofilled polymers. *Nat Commun.* **2016**, *7*, 1–6. [CrossRef] [PubMed]

34. Khare, H.S.; Burris, D.L. A quantitative method for measuring nanocomposite dispersion. *Polymer* **2010**, *51*, 719–729. [CrossRef]

35. Klüppel, M.; Schuster, R.H.; Heinrich, G. Structure and Properties of Reinforcing Fractal Filler Networks in Elastomers. *Rubber Chem. Technol.* **1997**, *70*, 243–255. [CrossRef]

36. Flory, P.J. Molecular Size Distribution in Three Dimensional Polymers. I. Gelation. *J. Am. Chem. Soc.* **1941**, *63*, 3083–3090. [CrossRef]

37. Caló, E.; Khutoryanskiy, V.V. Biomedical applications of hydrogels: A review of patents and commercial products. *Eur. Polym. J.* **2015**, *65*, 252–267. [CrossRef]

38. Gangopadhyay, R.; De, A. Conducting polymer nanocomposites: A brief overview. *Chem. Mater.* **2000**, *12*, 608–622. [CrossRef]

39. Min, C.; Zou, X.; Yang, Q.; Liao, L.; Zhou, G.; Liu, L. Near-infrared light responsive polymeric nanocomposites for cancer therapy. *Curr. Top. Med. Chem.* **2017**, *17*, 1805–1814. [CrossRef] [PubMed]

40. Li, S.; Meng Lin, M.; Toprak, M.S.; Kim, D.K.; Muhammed, M. Nanocomposites of polymer and inorganic nanoparticles for optical and magnetic applications. *Nano Rev.* **2010**, *1*, 1–19. [CrossRef] [PubMed]

41. Zaragoza, J.; Chang, A.; Asuri, P. Effect of crosslinker length on the elastic and compression modulus of poly(acrylamide) nanocomposite hydrogels. *J. Phys. Conf. Ser.* **2017**, *790*, 1–6. [CrossRef]

Article

Characterization of Nanoparticle Intestinal Transport Using an In Vitro Co-Culture Model

Alina F.G. Strugari [1], Miruna S. Stan [1,*], Sami Gharbia [2], Anca Hermenean [2,3] and Anca Dinischiotu [1]

[1] University of Bucharest, Faculty of Biology, Department of Biochemistry and Molecular Biology, 91-95 Splaiul Independentei, 050095 Bucharest, Romania; alina.strugari@gmail.com (A.F.G.S.); ancadinischiotu@yahoo.com (A.D.)

[2] Institute of Life Sciences, Vasile Goldis Western University of Arad, 86 Rebreanu, 310414 Arad, Romania; samithgh2@hotmail.com (S.G.); anca.hermenean@gmail.com (A.H.)

[3] Department of Histology, Faculty of Medicine, Pharmacy and Dentistry, Vasile Goldis Western University of Arad, 1 Feleacului, 310396 Arad, Romania

* Correspondence: miruna.stan@bio.unibuc.ro; Tel.: +40-21-318-1575

Received: 19 November 2018; Accepted: 17 December 2018; Published: 21 December 2018

Abstract: We aimed to obtain a tunable intestinal model and study the transport of different types of nanoparticles. Caco-2/HT29-MTX co-cultures of different seeding ratios (7:3 and 5:5), cultured on Transwell®systems, were exposed to non-cytotoxic concentration levels (20 μg/mL) of silicon quantum dots and iron oxide (α-Fe_2O_3) nanoparticles. Transepithelial electric resistance was measured before and after exposure, and permeability was assessed via the paracellular marker Lucifer Yellow. At regular intervals during the 3 h transport study, samples were collected from the basolateral compartments for the detection and quantitative testing of nanoparticles. Cell morphology characterization was done using phalloidin-FITC/DAPI labeling, and Alcian Blue/eosin staining was performed on insert cross-sections in order to compare the intestinal models and evaluate the production of mucins. Morphological alterations of the Caco-2/HT29-MTX (7:3 ratio) co-cultures were observed at the end of the transport study compared with the controls. The nanoparticle suspensions tested did not diffuse across the intestinal model and were not detected in the receiving compartments, probably due to their tendency to precipitate at the monolayer surface level and form visible aggregates. These preliminary results indicate the need for further nanoparticle functionalization in order to appropriately assess intestinal absorption in vitro.

Keywords: co-culture intestinal model; Caco-2; HT29-MTX; nanoparticle transport; quantum dots; iron oxide nanoparticles

---

1. Introduction

Oral drug administration is the preferred route when it comes to delivering most active compounds, with advantages including patient comfort, reduced chances of infection, and minimal invasiveness compared with alternative delivery systems. The assimilation process via the gastrointestinal (GI) tract is a paramount condition for the delivery of any xenobiotic active compound to target tissues. In order to reach the vascular circulation, orally administered drugs usually have to pass through the small intestinal barrier. However, some active compounds are especially vulnerable to the harsh GI environment and, therefore, require protection during transit in order to prevent degradation [1]. In this context, nanoparticles constitute novel candidates as future carrier-type agents [2]. Furthermore, human exposure to food products containing nanoparticulate materials is projected to increase in the near future, which calls for the development of a reliable screening solution [3].

The intestinal mucosa constitutes the major absorption site in vivo, with nutrient and xenobiotics having to penetrate two types of barriers—an acellular mucus layer and the *intestinal epithelium*. The latter is composed of a heterogeneous population of terminally differentiated cells which are constantly renewed by Lgr5+ and +4 stem cells located at the crypt base level, making the mammalian intestinal epithelium renowned for having one of the highest turnover rates in the body [4]. Up to 80% of the cells that make up the tissue are absorptive enterocytes, followed by mucus-secreting goblet cells which, ratio-wise, vary between 10 and 24%, depending on the GI tract region in which they are located [5]. Lymphatic follicles cluster in different regions of the ileum and are known as Peyer's patches. These patches are covered by a specialized tissue called the follicle-associated epithelium (FAE), which is composed of mainly enterocytes and M (microfold) cells that interact with the local immune system, mediating antigenic influx as well as bacterial attachment [2,6,7]. Paneth immune cells are also part of the epithelium—located at the base of Lieberkühn's crypts, they produce acidophilic granules containing antibacterial and digestive enzymes. Enteroendocrine cells account for around 1% of the intestinal monolayer and are sporadically dispersed throughout the intestinal mucosa [4].

There is currently a wide array of in vitro intestinal models being used in studies that aim to estimate/predict drug absorption in vivo. Even though there has been a surge in the development of organotypic (3D) models [8–10] and organs-on-chip (body-on-a-chip) systems [11,12], the majority of transport studies rely on simpler, in vitro co-culture models using conventional cell lines. There are many advantages in terms of cost and their good reproducibility and fidelity, yet the use of tumoral cell lines (which is common practice in most 2D in vitro intestinal models) raises several concerns regarding the ability of the models to accurately reflect in vivo intestinal absorption. More often than not, tumoral cells are found to overexpress key proteins [13,14], and they generally exhibit an altered transcriptional regulation phenotype that may impact tissue permeability. For example, the Caco-2 adenocarcinoma cell line has been used extensively for the past couple of decades in nutrient and drug transport studies as an adequate in vitro model of the intestinal mucosa [15,16]. However, due to the overexpression of tight junction protein complexes [5,17–20], simple Caco-2 monolayers fail to provide a reliable estimation of in vivo paracellular permeability of small hydrophilic compounds. To address this issue, Caco-2 cells are routinely co-cultured alongside HT29-MTX (goblet-like) cells [18] on Transwell® inserts (Figure 1).

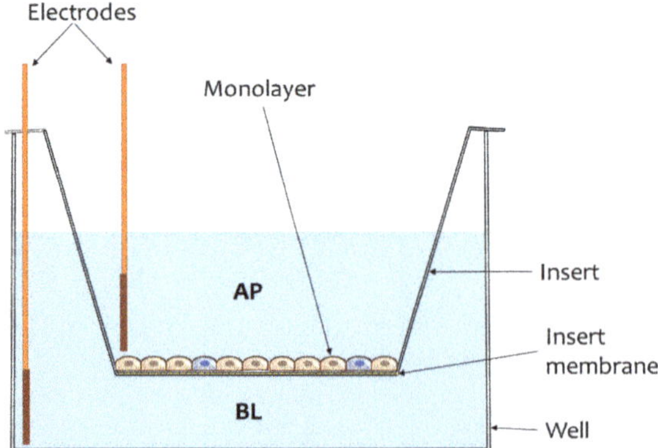

Figure 1. Illustration of a common experimental model used in intestinal transport studies. The cells are cultivated on inserts supplied with semipermeable membranes through which microparticulate diffusion may occur. Monolayer integrity is usually assessed by measuring transepithelial electrical resistance (TEER) using chopstick electrodes inserted in the apical (AP) and basolateral (BL) compartments of the well.

This co-culture model allows for modulating intercellular junction geometry, thereby fine-tuning the *effective permeability* (P_{eff}) of the monolayer by simply adjusting the initial cell seeding ratio [17].

The human adenocarcinoma line HT-29 preconditioned in methotrexate (MTX) has the added benefit of expressing mucins in culture [21]—the resultant mucus layer produced constitutes an additional physical barrier [6], potentially impeding xenobiotic transport across the epithelium as would be the case under in vivo conditions. Intestinal permeability correlates with the rate of compound transport across the mucosa, which is calculated according to the following equation:

$$P_{\text{eff}} = \frac{dQ}{dt} \frac{V}{A\,C_0}\ [\text{cm/s}],\tag{1}$$

where dQ/dt represents the apparent flow rate in time across the monolayer (mM/mL·s), V is the volume (mL) within the receiving compartment (BL), C_0 is the initial concentration of the compound (mM) in the donor compartment (AP), and A is the exposed tissue surface area (cm^2).

The physicochemical properties of the compounds that are tested in transport studies also have a huge impact on the rate of absorption and, in some cases, require functionalization in order to facilitate (or even allow) assimilation across the intestinal mucosa [3]. Depending on the nature of the compound, it may reach the other side of the epithelium by way of the transcellular or paracellular route (via passive diffusion), transcytosis, or active transport (via carrier-mediated transport), as depicted in Figure 2.

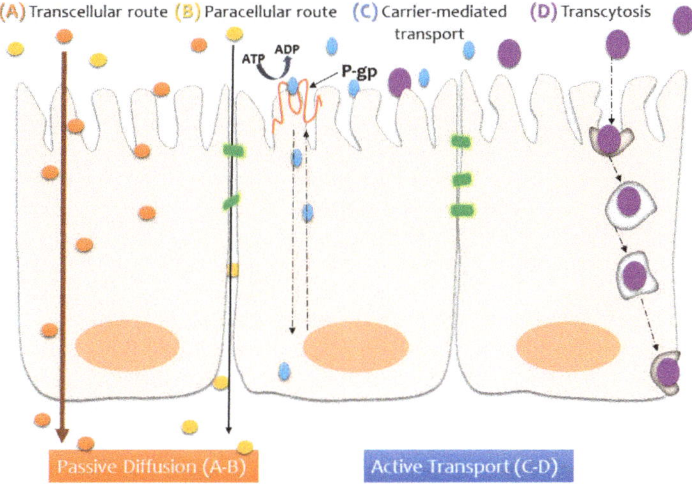

Figure 2. Schematic illustration of the possible transport pathways across the intestinal epithelium, depending on the nature of the compound being absorbed.

It is worth keeping in mind that, in reality, xenobiotic assimilation depends on many more variables. Bioavailability (F) is among the most crucial parameters in the field of pharmacokinetics, as it describes the fraction of compounds that manage to penetrate the intestinal barrier and reach blood circulation in their unaltered form [22]:

$$F = f_a\,(1 - E_G) \cdot (1 - E_H),\tag{2}$$

where f_a is the absorbed fraction of the dose administrated (mass/dose), while considering first-pass metabolism of the compound in the gut wall (E_G) and liver (E_H). These are variables which current in vitro co-culture models cannot account for.

Although Caco-2/HT29-MTX co-cultures are routinely used in transport studies, the model is yet to be fully characterized, especially when compared with the well-established Caco-2 monoculture model. The present study aims to add to the already established body of work in this direction [5,23] while assessing the potential of two types of nanoparticles for oral drug delivery or screening perspectives. We established several Caco-2/HT29-MTX cultures by altering the initial seeding ratios (10:0, 7:3, 5:5, 0:10). The cells were cultured using Transwell® systems and were allowed to develop into stable monolayers for 21 days before exposing them to non-cytotoxic concentrations (20 µg/mL) of silicon quantum dots (Si QDs) and iron oxide (α-Fe$_2$O$_3$) nanoparticles. According to our previous investigations [24,25], we chose to conduct the transport study using the concentration of 20 µg/mL for both types of nanoparticles. Those particular studies showed that this dose does not induce changes in the number of viable cells or have cytotoxic effects. Using a lower concentration would not be appropriate for the objective of the study, as it could impede proper particle detection by conventional methods and alter the particles' potential drug delivery function, while a higher concentration results in more aggregates and increased cell toxicity. Transepithelial electric resistance (TEER) was measured before and after exposure, and monolayer permeability (P_{eff}) was assessed via the paracellular marker Lucifer Yellow. At regular intervals during the 3-h transport study, samples were collected from the basolateral compartments for detection and quantitative testing. Cell morphology characterization was done by actin cytoskeleton labeling, and Alcian Blue/eosin staining was performed on insert cross-sections in order to compare the intestinal models and evaluate the production of mucins.

2. Materials and Methods

2.1. Nanoparticles

The Si QDs and α-Fe$_2$O$_3$ nanoparticles used in this study were produced at the Laser Department from the National Institute of Lasers, Plasma and Radiation Physics, Bucharest-Măgurele. The silicon nanoparticles were synthesized via a pulsed laser ablation technique by irradiating a silicon target in an ambient gas (such as argon or helium), which was introduced into the chamber at a constant pressure and gas flow rate. These nanoparticles exhibit a strong red visible luminescence, and the photoluminescence spectrum places the maximum intensity emission at ~690 nm. At room temperature, QDs show a broadband emission spectrum in the approximate range of 400–800 nm under a 325 nm excitation wavelength. The production process and characterization of the nanomaterials (Table 1) are detailed in previous works [24,26]. Briefly, TEM images reveal irregularly shaped nanoparticle particulate aggregates which can be easily disrupted and dispersed by sonication; the Si QDs have a spherical morphology and measure roughly 6–8 nm in diameter. Iron oxide nanoparticles of the α form of hematite (Fe$_2$O$_3$) were assessed using a high-resolution transmission electron microscope (Philips CM120). The size distribution places them in the range of 10–120 nm (most being between 40 and 60 nm). A Bruker AXS/D8 Advance X-ray diffractometer was used to conduct crystallinity analysis [25]. The characterization of hydrodynamic size and zeta potential was performed in distilled water using a Malvern Zetasizer Nano-ZS instrument (Malvern Instruments, Malvern, UK) with an equilibration time of 1 min at 25 °C using a refractive index of 1.52 for silicon dioxide and 2.93 for hematite.

Table 1. Physicochemical characteristics of α-Fe$_2$O$_3$ and Si quantum dot (QD) nanoparticles as described previously [24–26].

Characteristics	α-Fe$_2$O$_3$ [25]	Si QDs [24,26]
Form of synthesis	Powder	Powder
Method of synthesis	Laser ablation	Laser ablation
Purity	>99%	>99%
Structure	α form of Fe$_2$O$_3$ (hematite)	Crystalline silicon core covered by an amorphous SiO$_2$ shell
Morphology	Spherical particles	Spherical particles
Average particle size	40–60 nm	6–8 nm
Hydrodynamic size in water	250–300 nm	250–300 nm
Z-potential in water	−30 mV	−27 mV

2.2. Cell Culture

In a preliminary phase, Caco-2 (cat. no. CRL-2102) and HT-29 (cat. No. HTB-38) cell lines purchased from American Type Cell Culture (ATCC, USA) were grown separately in complete medium consisting of Dulbecco's Modified Eagle Medium (DMEM) supplemented with 10% heat-inactivated FBS and 1% antibiotics. The cultures were maintained in a humid atmosphere at 37 °C with 5% CO_2 and were routinely subcultured once a week with 0.25% trypsin and 0.53 mM EDTA. For several months, HT-29 cells were treated with methotrexate (MTX) according to the original protocol developed by Lesuffleur et al. [27].

Subsequent to the stabilization of HT29-MTX (mucus-secreting) clones, co-cultures were initiated. The cells were seeded on 12-well plates with Transwell® inserts (with polycarbonate membranes, 3 μm pore size, Corning B. V. Life Sciences, USA) at a final density of 100,000 cells per well, regardless of the final seeding ratio (Caco2:HT29-MTX): 10:0, 7:3, 5:5, and 0:10.

2.3. Transport Study Design

We exposed Caco-2, HT29-MTX, and co-cultures to non-cytotoxic concentrations (20 μg/mL) of Si QDs and α-Fe_2O_3 nanoparticles and to Lucifer Yellow (50 ug/mL). The nanoparticle suspensions were sonicated and dispersed using an ultrasonic processor Hielscher UP50H (Hielscher Ultrasonics GmbH, Germany). Hank's Balanced Salt Solution (HBSS) was chosen as the transport buffer during the experiment. Particles were quantified by measuring absorbance/fluorescence levels for each (α-Fe_2O_3 nanoparticles: absorbance −325/500 nm; Si QDs: wavelength excitation/emission −325/644 nm; wavelength excitation/emission −405/535 nm) using the Flex Station 3 Multireader (Molecular Devices, USA). TEER monitoring was performed using a Millipore®Millicell Electrical Resistance (ERS) system (Millipore, USA). Measurements were performed at three different points of each well, and the final TEER was calculated according to the following formula:

$$\text{TEER}_{\text{final}} = (\text{TEER}_{\text{mean}}\ [\Omega] - \text{TEER}_{\text{blank}}\ [\Omega]) \times A_{\text{well}}\ (1.12\ \text{cm}^2)\ [\Omega\ \text{cm}^2] \tag{3}$$

2.4. Cell Morphology Analysis

At the end of an incubation time of 3 h, Transwell inserts were removed and fixed in 4% paraformaldehyde (PFA) for 24 h before being paraffinized, sectioned (5 μm thick cross-sections), and stained with 1% Alcian Blue-8GX and 0.1% eosin. The procedure was done according to the manufacturer's kit (Bio-Optica) instructions. Briefly, the sections were placed in distilled water. Alcian Blue pH 2.5 reagent was left to react for 30 min, after which the slides were drained and exposed to sodium tetraborate solution for 10 min. The slides were washed in distilled water before and after treatment with Carmalum reagent (5 min). After a series of dehydration steps using ascending alcohols and xylene, the sections were mounted, and images were captured using an Olympus BX43 (XC30 software). Other cells that were cultured on inserts were fixed with 4% paraformaldehyde for 20 min and permeabilized with 0.1% Triton X-100 and 2% bovine serum albumin for 40 min. F-actin was stained for 30 min with 10 μg/mL phalloidin-FITC (fluorescein isothiocyanate), and the nuclei were counterstained with 2 μg/mL DAPI (4′,6-diamino-2-phenylindole). The slides were examined with an inverted fluorescence microscope Olympus IX71 (Olympus, Tokyo, Japan).

2.5. Statistical Analysis

All experiments were performed in triplicate. Data are expressed as the mean value ± standard deviation (SD) of the three independent experiments (n = 3). Statistical differences between samples were analyzed by Student's *t*-test executed in Microsoft Office software (Excel 2007), and a value of $p < 0.05$ was considered to be statistically significant.

3. Results

3.1. Monolayer Integrity Assessment

The intestinal models tested in the present study recorded similar TEER values to those reported in the literature. After 14 days in culture, the Caco-2 cell line alone produced a very compact monolayer (TEER$_\mu$ = 369 Ω cm^2); measurements taken a week after revealed a steep drop to 228 Ω cm^2, and a similar phenomenon was observed in the case of the Caco-2/HT29-MTX co-cultures seeded at a 7:3 ratio. This would indicate a tissular integrity alteration of around 38%, as seen in Figure 3. The co-cultures initially seeded at equal ratios and goblet cell-like monocultures evolved in a predictable pattern, with steadily increasing TEER values over time. However, the TEER measurements taken 1 week apart for the co-cultures with a higher ratio of Caco-2 (7:3), as well as the Caco-2 monolayers alone, were lower than expected. Nonetheless, after 21 days, the TEER values overall maintained a similar differential rate of increase across models, corresponding to the increasing ratio of Caco-2 initially seeded.

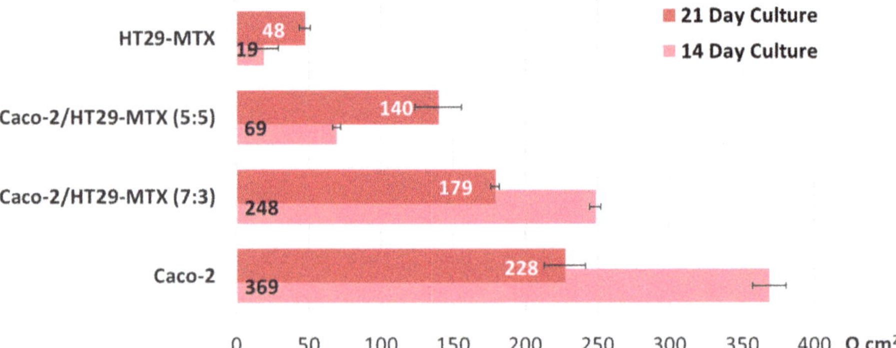

Figure 3. Transepithelial electric resistance (TEER) evolution before initiating the transport study. The TEER values were measured 14 and 21 days after seeding (on Transwell inserts) the Caco-2:HT29-MTX cell cultures in four different ratios: 0:1, 5:5, 7:3, and 1:0. Data are expressed as means $\pm$ standard deviation (SD) (n = 3).

Regardless of whether the models were exposed to the nanoparticle suspension or not, at the end of the experiment, all groups displayed significantly lower TEER values (see Figure 4), which suggests that other variables at play affect monolayer integrity. Interestingly, when compared with the monocultures, the co-culture models recorded the highest alteration levels, but only if they were exposed to either type of nanoparticle suspension during this time. The 7:3 seeding ratio (Caco-2/HT29-MTX) was selected for further investigations, as it is a better model in terms of approaching the in vivo ratio of enterocytes and mucus-producing goblet cells compared with the other ratios tested. As seen in Figure 3, the Caco-2/HT29-MTX model seeded at a ratio of 7:3 was able to achieve TEER values similar to those cited in the literature [5,28,29], which points to an adequate distribution of tight junctions and confluent cell layer homogeneously covered with mucus.

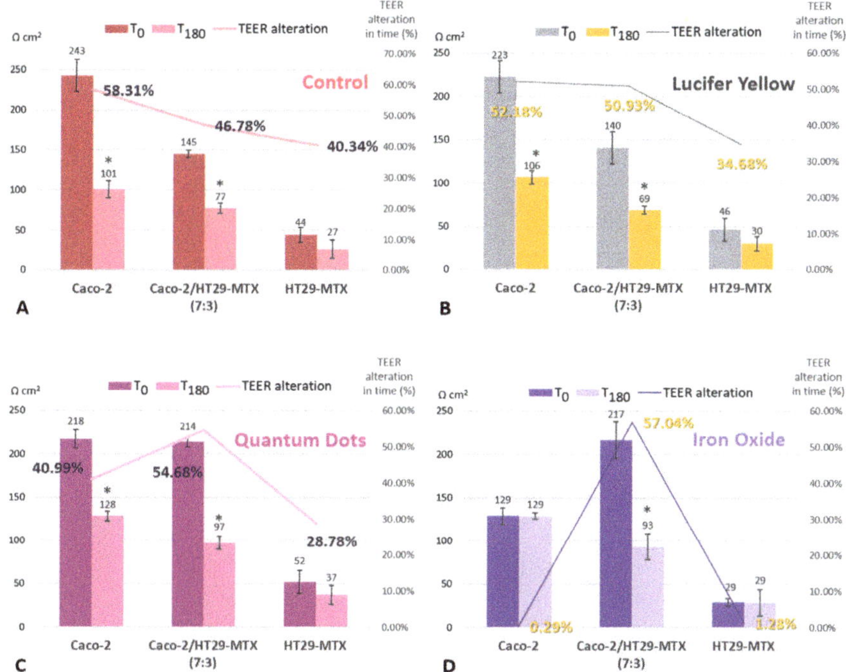

Figure 4. TEER alteration for control cells (**A**), cells incubated with Lucifer Yellow (**B**), with Si QDs (**C**), and with α-Fe$_2$O$_3$ nanoparticles (**D**). Comparison between three intestinal models (Caco-2, HT29-MTX, and co-culture) before initiating the transport study (T$_0$) and after a 3-h exposure to the nanoparticles (T$_{180}$). TEER alteration in time (expressed as a percentage) was calculated in order to compare the difference between the initial values and those measured at 180 min. The calculation was as follows: TEER alteration percent (%) = (T$_0$ − T$_{180}$)/T$_0$ × 100. Data are expressed as means ± standard deviation (SD) (n = 3). * $p < 0.05$ compared with T$_0$.

3.2. Comparative Morphological Characterization

In order to assess whether the HT29-MTX cell line had produced mucus, three inserts were chosen from the control group, each containing a different seeding-ratio developed model, as seen in Figure 5.

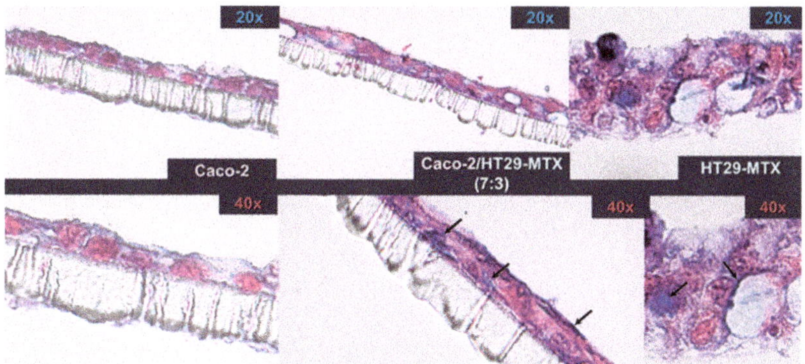

Figure 5. Insert cross-sections after 21 days in culture. Staining with 1% Alcian Blue highlights the production of mucins (black arrows) by HT29-MTX cells. The monolayers were counterstained with 0.1% eosin.

Despite the fact that all co-culture models had an identical initial seeding density, Alcian Blue-stained cross-sections show that, after 3 weeks, HT29-MTX monocultures proliferated more, resulting in the formation of a 3D diffused multilayer around the polycarbonate membrane. Compared with the Caco-2 monolayers, the co-culture model exhibited a more intense Alcian Blue staining, which suggests the incorporation of HT29-MTX mucus-producing cells.

Alcian Blue staining revealed morphological alterations following the 3-h exposure to Si-QDs or Fe_2O_3 nanoparticle suspensions (Figure 6). Comparative characterization of the monolayers after exposure was also conducted using F-actin staining (Figure 7), and the results corroborated the TEER results and the cross-sectional areas stained with Alcian Blue.

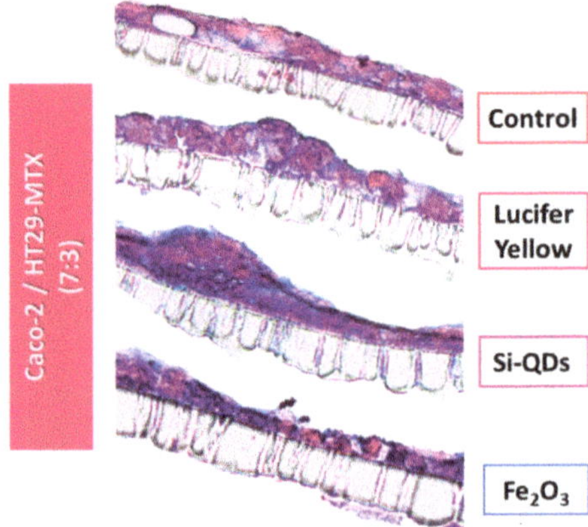

Figure 6. Insert cross-sections of the Caco-2/HT29-MTX intestinal model seeded at a 7:3 ratio following a 3-h exposure to nanoparticles. The cells were stained with 1% Alcian Blue/0.1% eosin in order to detect mucin production. Monolayer disruptions of the experimental groups are visible compared with controls.

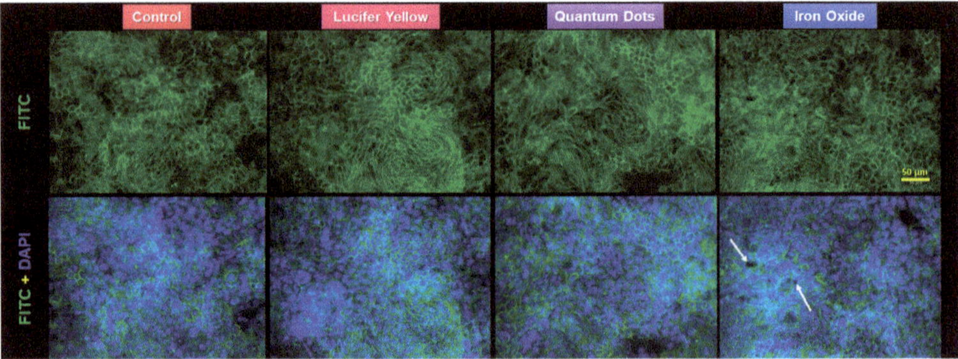

Figure 7. Fluorescence staining of the actin cytoskeleton of Caco-2/HT29-MTX monolayer after the 3-h transport study. F-actin is marked with fluorescein isothiocyanate (FITC)-phalloidin (green). Nuclear counterstain with 4′,6-diamidino-2-phenylindole (DAPI) (blue). Note the aggregates of α-Fe_2O_3 nanoparticles (*white arrows*) on the top of the cells.

3.3. Transport Study

During the 3-h transport study, neither Si QDs nor Fe_2O_3 nanoparticles permeated across the monolayer insert to reach the receiving compartment (data not shown). The paracellular marker Lucifer Yellow passively diffused and reached the basolateral compartment, where it was detected

(Figure 8). The Lucifer Yellow transport rate correlates with TEER measurements—the progressive integrity loss of the monolayers and, implicitly, the loosening of tight junction complexes established by Caco-2 cells would result in the increase of effective permeability, even without considering their initial seeding ratio.

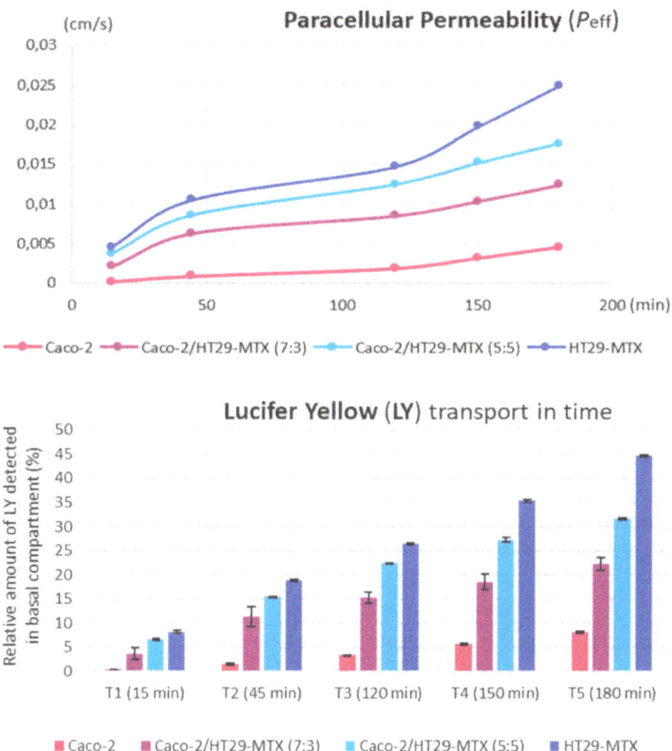

Figure 8. Effective permeability increases across all intestinal in vitro variants (top graph). Lucifer Yellow transport over time (bottom graph) shows the relative amount (in percentages) of the paracellular marker, which was detected at five measurement points during the experiment. Data are calculated as means ± standard deviation (SD) (n = 3).

4. Discussion

Drug permeability coefficients have been historically assessed using simple Caco-2 cultures. This classical intestinal model does present itself with many limitations, one of which is an abnormally high transepithelial electrical resistance (TEER). TEER measurements ensure the consistent monitoring of tissular integrity because it is a highly sensitive, non-invasive technique. TEER values also inversely correlate with paracellular permeability, making it the method of choice for intestinal transport studies [30]. In vivo studies reveal different TEER values between different regions of the human gut; post-confluent Caco-2 monocultures are known to generate signals that vary between 150 and 500 Ω cm^2, whereas in vivo TEER recordings place the intestinal epithelium in a realistic range of 12–69 Ω cm^2 [17,30]. This significant difference can be explained by the fact that Caco-2 cells in culture form many more tight junctions.

When comparing the TEER measurements of co-cultures seeded at different cell ratios (5:5 and 7:3), the data imply convergence in time toward similar transepithelial electrical resistance values (Figure 3). This suggests an internal restructuring process which alters the initial cell type seeding ratio and is congruent with the work of Chen et al. [28], whose study showed, using a Taguchi experimental design, that neither the initial enterocyte-goblet cell ratio nor the cellular density had a significant impact on TEER recordings or the transport of active compounds. Intercellular signaling could account for these results, with both cell lines modulating each other's growth and proliferation and reaching an equilibrium, regardless of the initial seeding rate. Tumoral cell lines are known to produce and release a series of growth factors which play a crucial role in intestinal development [31]. For example, the expression of epidermal growth factor receptors (EGFRs) in Caco-2 cells is modulated by the substrate on which they are grown, as well as their differentiation stage [32]. Their cellular proliferation is highly affected if they are exposed to vitamin D_3, which also leads to increases in alkaline phosphatase (ALP) expression [33]. In contrast, insulin-like growth factor 1 (IGF-1) has a positive, dose-dependent effect on Caco-2 proliferation [31].

TEER measurements taken 1 week later for the co-cultures with the higher ratio of Caco-2 (7:3) and the Caco-2 monolayers alone were lower than expected (Figure 3). One possible explanation for this is that, unlike HT29-MTX, the Caco-2 cell line has higher requirements in terms of cell culture media, thereby inhibiting cellular expansion in a post-confluent setting. Caco-2 cells grow well on media enriched with 16.5% fetal calf serum (FCS); however, co-culture models must consider the requirements of all the cell lines involved. This study took into account recommendations provided by other research [5] and exposed the cells to a lower concentration of fetal bovine serum (10% FBS). Unlike Caco-2, goblet-like HT29-MTX cells grow well in less strict conditions and require a low amount of glucose; moreover, the cell line is a known producer of lactic acid which can significantly lower the environmental pH. This was observed during handling of the cell line and may account for the lower than expected TEER values of the co-cultures [27].

Previous studies show that many factors, including medium composition and cell culture time, impact TEER values and permeability coefficients of many routinely tested drugs [5,31]. This points to the degradation of the monolayers after a certain time in culture elapses, which results in more permeable intestinal models for compounds that employ the paracellular route. This hypothesis, however, does not account for the higher alteration levels observed for the co-culture models that were part of the experimental groups. This preliminary result suggests that even though the nanoparticle suspensions were dosed such as to not reach cytotoxic levels for either cell line, the co-culture model somehow renders them more vulnerable and more responsive to the materials.

One possible explanation was given by Soto et al. [34], who attributed the cytotoxic effects of some nanoparticulate materials to their inherent tendency to precipitate on top of the cells and form visible aggregates. We noticed a similar phenomenon occurring, prompting further work in order to properly functionalize them and increase their solubility in buffer solutions. Iron oxide nanoparticles are often challenging to work with in this regard—for example, cellular intake of superparamagnetic iron oxide nanoparticle variants (UPSIO NPs) required further functionalization with an oleic acid coating [35]. Even though iron oxide particles are insoluble at physiological temperature and pH, interfacial hydration of anhydrous hematite increases solubility, which enables the cellular uptake of α-Fe_2O_3 NPs with the release of ferric ions and generation of hydroperoxyl radicals [25].

QD uptake via the gastrointestinal tract is a new subject that has attracted researchers' attention due to their biomedical applications, such as in vivo imaging and drug delivery. Degradation of QDs can occur in acidic environments, including the stomach, and can release toxic elements, such as cadmium ions in the case of CdSe-based QDs. Thus, the evaluation of the biodistribution and stability of QDs in the digestive tract is a matter of great interest nowadays. Studies have shown the possibility to modify the surface of particles by different coating approaches in order to achieve their resistance to strong acidic solutions and intestinal permeability [36]. Due to their tendency to aggregate [24], QDs will not easily translocate through the intestinal epithelial barrier [37], as we have seen in this current study.

Nanoparticle transport across the monolayer was not detected using the quantitative method described in this study. Our working hypothesis is that both types of nanoparticles tested could not diffuse across the monolayer due to their tendency to form aggregates and settle on top of the cells. These nanoparticle aggregates are of a micrometric size and could not be measured by conventional, currently used methods, such as dynamic light scattering (DLS), as this assay requires a properly prepared sample. We measured the hydrodynamic size by DLS (Table 1) of the samples without aggregates, and we obtained values between 250 and 300 nm regardless of the type of particles (quantum dots or iron oxide particles), with the values increasing in the cell culture medium, as previously shown [38]. In addition, the Z-potential of the tested nanoparticles was negative in distilled water, near the value of −30 mV, confirming the negative charge of these particles. We suggest that this Z-potential value did not interfere with the TEER measurements, as we did not observe any difference between the value measured in the Transwell inserts with only PBS (blank) and those only with particles in PBS.

In conclusion, the current study partly achieved its goals concerning the characterization of the in vitro intestinal model Caco-2/HT29-MTX in a transport study setting. Further quantitative testing for the purpose of evaluating nanoparticle toxicity is required, as well as preventing aggregation.

Author Contributions: Conceptualization, M.S.S. and A.D.; Investigation, A.F.G.S., M.S.S., S.G. and A.H.; Methodology, A.F.G.S. and S.G.; Project administration, M.S.S.; Supervision, M.S.S., A.H. and A.D.; Writing—original draft, A.F.G.S.; Writing—review & editing, M.S.S., A.H. and A.D.

Funding: This research was funded by UEFISCDI agency through the project no. 77/2018 NANO-BIO-INT.

Acknowledgments: We are very grateful to Cristina Nistor from ICECHIM for providing access to Malvern instrument.

Conflicts of Interest: The authors declare no conflict of interest.

References

1. Wan, Z.-L.; Guo, J.; Yang, X.-Q. Plant protein-based delivery systems for bioactive ingredients in foods. *Food Funct.* **2015**, *6*, 2876–2889. [CrossRef] [PubMed]
2. Schimpel, C.; Teubl, B.; Absenger, M.; Meindl, C.; Fröhlich, E.; Leitinger, G.; Zimmer, A.; Roblegg, E. Development of an advanced intestinal in vitro triple culture permeability model to study transport of nanoparticles. *Mol. Pharm.* **2014**, *11*, 808–818. [CrossRef] [PubMed]
3. Walczak, A.P.; Kramer, E.; Hendriksen, P.J.M.; Tromp, P.; Helsper, J.P.F.G.; Van Der Zande, M.; Rietjens, I.M.C.M.; Bouwmeester, H. Translocation of differently sized and charged polystyrene nanoparticles in in vitro intestinal cell models of increasing complexity. *Nanotoxicology* **2015**, *9*, 453–461. [CrossRef]
4. van der Flier, L.G.; Clevers, H. Stem cells, self-renewal, and differentiation in the intestinal epithelium. *Annu. Rev. Physiol.* **2009**, *71*, 241–260. [CrossRef]
5. Hilgendorf, C.; Spahn-langguth, H.; Regårdh, C.G.; Lipka, E. Caco-2 versus Caco-2/HT29-MTX co-cultured cell lines: Permeabilities via diffusion, inside- and outside-directed carrier-mediated transport. *J. Pharm. Sci.* **2000**, *89*, 63–75. [CrossRef]
6. Lechanteur, A.; das Neves, J.; Sarmento, B. The role of mucus in cell-based models used to screen mucosal drug delivery. *Adv. Drug Deliv. Rev.* **2017**, *124*, 50–63. [CrossRef] [PubMed]
7. Kim, S.H.; Jang, Y.S. Antigen targeting to M cells for enhancing the efficacy of mucosal vaccines. *Exp. Mol. Med.* **2014**, *46*, e85. [CrossRef]
8. Chen, Y.; Lin, Y.; Davis, K.M.; Wang, Q.; Rnjak-Kovacina, J.; Li, C.; Isberg, R.R.; Kumamoto, C.A.; Mecsas, J.; Kaplan, D.L. Robust bioengineered 3D functional human intestinal epithelium. *Sci. Rep.* **2015**, *5*, 13708. [CrossRef]
9. Leushacke, M.; Barker, N. Ex vivo culture of the intestinal epithelium: Strategies and applications. *Gut* **2014**, *63*, 1345–1354. [CrossRef]
10. Costello, C.M.; Hongpeng, J.; Shaffiey, S.; Yu, J.; Jain, N.K.; Hackam, D.; March, J.C. Synthetic small intestinal scaffolds for improved studies of intestinal differentiation. *Biotechnol. Bioeng.* **2014**, *111*, 1222–1232. [CrossRef]
11. Sarmento, B.; Andrade, F.; da Silva, S.B.; Rodrigues, F.; das Neves, J.; Ferreira, D. Cell-based in vitro models for predicting drug permeability. *Expert Opin. Drug Metab. Toxicol.* **2012**, *8*, 607–621. [CrossRef] [PubMed]
12. Sakolish, C.M.; Esch, M.B.; Hickman, J.J.; Shuler, M.L.; Mahler, G.J. Modeling barrier tissues in vitro: Methods, achievements, and challenges. *EBioMedicine* **2016**, *5*, 30–39. [CrossRef] [PubMed]

13. Fanning, A.S.; Van Itallie, C.M.; Anderson, J.M. Zonula occludens-1 and -2 regulate apical cell structure and the zonula adherens cytoskeleton in polarized epithelia. *Mol. Biol. Cell* **2012**, *23*, 577–590. [CrossRef] [PubMed]

14. Selga, E.; Noé, V.; Ciudad, C.J. Transcriptional regulation of aldo-keto reductase 1C1 in HT29 human colon cancer cells resistant to methotrexate: Role in the cell cycle and apoptosis. *Biochem. Pharmacol.* **2008**, *75*, 414–426. [CrossRef] [PubMed]

15. Artursson, P.; Palm, K.; Luthman, K. Caco-2 monolayers in experimental and theoretical predictions of drug transport. *Adv. Drug Deliv. Rev.* **2012**, *64*, 280–289. [CrossRef]

16. Lechanteur, A.; Almeida, A.; Sarmento, B. Elucidation of the impact of cell culture conditions of Caco-2 cell monolayer on barrier integrity and intestinal permeability. *Eur. J. Pharm. Biopharm.* **2017**, *119*, 137–141. [CrossRef] [PubMed]

17. Béduneau, A.; Tempesta, C.; Fimbel, S.; Pellequer, Y.; Jannin, V.; Demarne, F.; Lamprecht, A. A tunable Caco-2/HT29-MTX co-culture model mimicking variable permeabilities of the human intestine obtained by an original seeding procedure. *Eur. J. Pharm. Biopharm.* **2014**, *87*, 290–298. [CrossRef] [PubMed]

18. Walter, E.; Janich, S.; Roessler, B.J.; Hilfinger, J.M.; Amidon, G.L. HT29-MTX/Caco-2 cocultures as an in vitro model for the intestinal epithelium: In vitro-in vivo correlation with permeability data from rats and humans. *J. Pharm. Sci.* **1996**, *85*, 1070–1076. [CrossRef] [PubMed]

19. Artursson, P.; Karlsson, J. Correlation between oral drug absorption in humans and apparent drug permeability coefficients in human intestinal epithelial (Caco-2) cells. *Biochem. Biophys. Res. Commun.* **1991**, *175*, 880–885. [CrossRef]

20. Van Itallie, C.M.; Anderson, J.M. Architecture of tight junctions and principles of molecular composition. *Semin. Cell Dev. Biol.* **2014**, *36*, 157–165. [CrossRef]

21. Kitamura, H.; Cho, M.; Lee, B.H.; Gum, J.R.; Siddiki, B.B.; Ho, S.B.; Toribara, N.W.; Lesuffleur, T.; Zweibaum, A.; Kitamura, Y.; et al. Alteration in mucin gene expression and biological properties of HT29 colon cancer cell subpopulations. *Eur. J. Cancer* **1996**, *32*, 1788–1796. [CrossRef]

22. Lennernäs, H. Intestinal permeability and its relevance for absorption and elimination. *Xenobiotica* **2007**, *37*, 1015–1051. [CrossRef] [PubMed]

23. Mahler, G.J.; Shuler, M.L.; Glahn, R.P. Characterization of Caco-2 and HT29-MTX cocultures in an in vitro digestion/cell culture model used to predict iron bioavailability. *J. Nutr. Biochem.* **2009**, *20*, 494–502. [CrossRef] [PubMed]

24. Stan, M.S.; Memet, I.; Sima, C.; Popescu, T.; Teodorescu, V.S.; Hermenean, A.; Dinischiotu, A. Si/SiO$_2$ quantum dots cause cytotoxicity in lung cells through redox homeostasis imbalance. *Chem. Biol. Interact.* **2014**, *220*, 102–115. [CrossRef] [PubMed]

25. Radu, M.; Munteanu, M.C.; Petrache, S.; Serban, A.I.; Dinu, D.; Hermenean, A.; Sima, C.; Dinischiotu, A. Depletion of intracellular glutathione and increased lipid peroxidation mediate cytotoxicity of hematite nanoparticles in MRC-5 cells. *Acta Biochim. Pol.* **2010**, *57*, 355–360. [PubMed]

26. Grigoriu, C.; Kuroki, Y.; Nicolae, I.; Zhu, X.; Hirai, M.; Suematsu, H.; Takata, M.; Yatsui, K.; Physics, R. Photo and cathodoluminescence of Si/SiO$_2$ nanoparticles produced by laser ablation. *J. Optoelectron. Adv. Mater.* **2005**, *7*, 2979–2984.

27. Lesuffleur, T.; Barbat, A.; Dussaulx, E.; Zweibaum, A. Growth adaptation to methotrexate of HT-29 human colon carcinoma cells is associated with their ability to differentiate into columnar absorptive and mucus-secreting cells. *Cancer Res.* **1990**, *50*, 6334–6343.

28. Pham, V.T.; Seifert, N.; Richard, N.; Raederstorff, D.; Steinert, R.E.; Prudence, K.; Mohajeri, M.H. The effects of fermentation products of prebiotic fibres on gut barrier and immune functions in vitro. *PeerJ* **2018**, *6*, e5288. [CrossRef]

29. Chen, X.M.; Elisia, I.; Kitts, D.D. Defining conditions for the co-culture of Caco-2 and HT29-MTX cells using Taguchi design. *J. Pharmacol. Toxicol. Methods.* **2010**, *61*, 334–342. [CrossRef]

30. Srinivasan, B.; Kolli, A.R.; Esch, M.B.; Abaci, H.E.; Shuler, M.L.; Hickman, J.J. TEER measurement techniques for in vitro barrier model systems. *J. Lab. Autom.* **2015**, *20*, 107–126. [CrossRef]

31. Sambuy, Y.; De Angelis, I.; Ranaldi, G.; Scarino, M.L.; Stammati, A.; Zucco, F. The Caco-2 cell line as a model of the intestinal barrier: Influence of cell and culture-related factors on Caco-2 cell functional characteristics. *Cell Biol. Toxicol.* **2005**, *21*, 1–26. [CrossRef] [PubMed]

32. Kuwada, S.K.; Li, X.F.; Damstrup, L.; Dempsey, P.J.; Coffey, R.J.; Wiley, H.S. The dynamic expression of the epidermal growth factor receptor and epidermal growth factor ligand family in a differentiating intestinal epithelial cell line. *Growth Factors* **1999**, *17*, 139–153. [CrossRef] [PubMed]

33. Giuliano, A.R.; Franceschi, R.T.; Wood, R.J. Characterization of the vitamin D receptor from the Caco-2 human colon carcinoma cell line: Effect of cellular differentiation. *Arch. Biochem. Biophys.* **1991**, *285*, 261–269. [CrossRef]

34. Soto, K.; Garza, K.M.; Murr, L.E. Cytotoxic effects of aggregated nanomaterials. *Acta Biomater.* **2007**, *3*, 351–358. [CrossRef] [PubMed]

35. Kenzaoui, B.H.; Vilà, M.R.; Miquel, J.M.; Cengelli, F.; Juillerat-Jeanneret, L. Evaluation of uptake and transport of cationic and anionic ultrasmall iron oxide nanoparticles by human colon cells. *Int. J. Nanomed.* **2012**, *7*, 1275–1286.

36. Loginova, Y.F.; Dezhurov, S.V.; Zherdeva, V.V.; Kazachkina, N.I.; Wakstein, M.S.; Wakstein, M.S.; Savitsky, A.P. Biodistribution and stability of CdSe core quantum dots in mouse digestive tract following per os administration: Advantages of double polymer/silica coated nanocrystals. *Biochem. Biophys. Res. Commun.* **2012**, *419*, 54–59. [CrossRef] [PubMed]

37. Peuschel, H.; Ruckelshausen, T.; Kiefer, S.; Silina, Y.; Kraegeloh, A. Penetration of CdSe/ZnS quantum dots into differentiated vs undifferentiated Caco-2 cells. *J. Nanobiotechnol.* **2016**, *14*, 70. [CrossRef] [PubMed]

38. Stan, M.S.; Sima, C.; Cinteza, L.O.; Dinischiotu, A. Silicon-based quantum dots induce inflammation in human lung cells and disrupt extracellular matrix homeostasis. *FEBS J.* **2015**, *282*, 2914–2929. [CrossRef] [PubMed]

 *nanomaterials*

Article

Synthesis and Characterization of Graphene Oxide Derivatives via Functionalization Reaction with Hexamethylene Diisocyanate

Jose Antonio Luceño-Sánchez [1], Georgiana Maties [2], Camino Gonzalez-Arellano [2] and Ana Maria Diez-Pascual [1,*]

[1] Departamento de Química Analítica, Química Física e Ingeniería Química, Facultad de Ciencias, University of Alcalá, E-28871 Madrid, Spain; jose.luceno@uah.es

[2] Departamento de Química Orgánica y Química Inorgánica, Facultad de Ciencias, University of Alcalá, E-28871 Madrid, Spain; georgiamaties@gmail.com (G.M.); camino.gonzalez@uah.es (C.G.-A.)

* Correspondence: am.diez@uah.es; Tel.: +34-918-856-430

Received: 28 September 2018; Accepted: 20 October 2018; Published: 23 October 2018

Abstract: Graphene oxide (GO), the oxidized form of graphene, shows unique properties including high mechanical strength, optical transparency, amphiphilicity and surface functionalization capability that make it attractive in fields ranging from medicine to optoelectronic devices and solar cells. However, its insolubility in non-polar and polar aprotic solvents hinders some applications. To solve this issue, novel functionalization strategies are pursued. In this regard, this study deals with the preparation and characterization of hexamethylene diisocyanate (HDI)-functionalized GO. Different reaction conditions were tested to optimize the functionalization degree (FD), and detailed characterizations were conducted via elemental analysis, Fourier-transformed infrared (FT-IR) and Raman spectroscopies to confirm the success of the functionalization reaction. The morphology of HDI-GO was investigated by transmission electron microscopy (TEM), which revealed an increase in the flake thickness with increasing FD. The HDI-GO showed a more hydrophobic nature than pristine GO and could be suspended in polar aprotic solvents such as *N,N*-dimethylformamide (DMF), *N*-methylpyrrolidone (NMP) and dimethyl sulfoxide (DMSO) as well as in low polar/non-polar solvents like tetrahydrofuran (THF), chloroform and toluene; further, the dispersibility improved upon increasing FD. Thermogravimetric analysis (TGA) confirmed that the covalent attachment of HDI greatly improves the thermal stability of GO, ascribed to the crosslinking between adjacent sheets, which is interesting for long-term electronics and electrothermal device applications. The HDI-GO samples can further react with organic molecules or polymers via the remaining oxygen groups, hence are ideal candidates as nanofillers for high-performance GO-based polymer nanocomposites.

Keywords: graphene oxide; functionalization; hexamethylene diisocyanate; dispersion; functionalization degree; morphology; hydrophobicity; thermal stability

---

1. Introduction

Graphene (G), an allotrope of carbon like diamond, graphite and fullerenes, has attracted a lot interest in recent years both for fundamental studies and potential applications [1]. It is a flat, atomically thick two-dimensional (2D) sheet composed of sp^2 carbon atoms arranged in a honeycomb structure. It presents superior electronic, thermal and mechanical properties, very large surface area and the highest electrical conductivity known at room temperature [2]. Its extremely high carrier mobility, broad absorption spectral range, high optical transparency and abundance make very attractive material in various fields ranging from medicine [3] or high-performance composites [4] to chemical sensors [5] and solar cells [6].

Several routes have been reported for the preparation of G, including chemical vapor deposition (CVD) of hydrocarbons onto transition metal surfaces, micromechanical exfoliation of graphite, epitaxial growth on electrically insulating substrates like SiC wafers, electrochemical intercalation, thermal exfoliation or chemical reduction of graphite oxide [7]. However, these approaches lead to a low production yield and are time consuming. Further, due to its hydrophobic nature and strong van der Waals forces between adjacent sheets, it is insoluble in water or common organic solvents and displays poor dispersion in most solvents, which hinders its applications. In this context, graphene oxide (GO), originated from the exfoliation of graphite oxide or the chemical oxidation of G [8], has been studied in much research as an alternative to G. The established advantages of GO in production yield and cost make it an attractive candidate as nanofiller in polymer composites. GO is a water-soluble nanomaterial since it comprises epoxide, hydroxyl and carbonyl groups on the basal planes and carboxylic acids on the edges. Thus, upon sonication in aqueous media, it easily exfoliates and forms stable colloidal suspensions due to its strong hydrophilicity [9]. However, the exfoliation of GO in organic solvents is hindered due to strong hydrogen bonding interactions between adjacent layers. Improving the dispersibility of G-based materials in organic media is crucial from an application viewpoint, hence novel functionalization strategies are pursued. In this regard, the dispersion ability of G and GO in a wide range of solvents has been reviewed [10,11]. If the number of hydrogen bond donor groups in GO is reduced via chemical functionalization, the layers would become less hydrophilic and the strength of interlayer hydrogen bonding will be attenuated, thus allowing for exfoliation in organic solvents. Such functionalization was first demonstrated by Lerf et al. [12], who prepared a series of chemically modified graphite oxide derivatives.

Owed to its π electrons delocalized over the entire 2D network, G is somewhat chemically inert, hence covalent chemical functionalization of pristine G is a challenging task that typically requires reactive species that can form covalent adducts with the sp^2 carbon structures in G through free radical addition, CH insertion, or cycloaddition reactions [13]. Consequently, most of the research activity has been carried out in the field of chemical modification of GO, one of the hottest topics in nanotechnology that has been explored by a large number of studies [14,15]. In particular, the addition of nucleophilic species, such as amines or hydroxyls, attaches functional groups to GO via formation of amides or esters [14]. The attachment of aromatic or aliphatic amines like polyaniline, pyrrolidine, ethylenediamine, and so forth led to GO derivatives that were homogeneously dispersed in water, *N*,*N*-dimethylformamide (DMF), dimethyl sulfoxide (DMSO), and ethanol, but dispersed poorly in methanol, acetone or propanol and could not be suspended in chloroform, dichloromethane or toluene [16,17]. Polymers such as poly(ethylene glycol), poly(vinyl alcohol) and poly(2-(dimethylamino)ethyl methacrylate) have also been attached to GO via grafting-to or grafting-from approaches [18,19], and the resulting materials showed increased dispersibility in many solvents, including water, methanol, NMP or DMSO.

An interesting approach is the isocyanate functionalization proposed by Stankovich et al. [20]. In their work, reactions between different organic isocyanates and the hydroxyl and carboxyl groups of GO were tested to reduce the amount of hydrogen bonds with donor groups on GO sheets, thus reducing the nanomaterial hydrophilicity. As a result, the isocyanate-treated GO samples could be exfoliated in some polar aprotic solvents such as DMF, after a mild ultrasonication, albeit they could not be dispersed in non-polar solvents. In addition, the epoxy groups of GO can be easily modified through ring-opening reactions, and the resulting materials were well-dispersed in solvents such as water, DMF and DMSO [21].

In most of the aforementioned studies, the resulting GO derivatives showed improved dispersibility both in aprotic and protic polar organic solvents. However, their dispersibility in moderately polar or non-polar solvents was not increased, and this still restricts their use in certain applications like polymeric-based solar cells. In this regard, the aim of the present work is to synthesize and characterize chemically modified GO samples via treatment with hexamethylene diisocyanate (HDI) that can form stable dispersions in a wide range of solvents with different polarities. For such purpose, GO was first prepared using a modified Hummers' method from flake graphite [8] and then reacted with HDI in the presence of triethylamine (TEA) as a catalyst to yield functionalized HDI-GO nanosheets (Scheme 1).

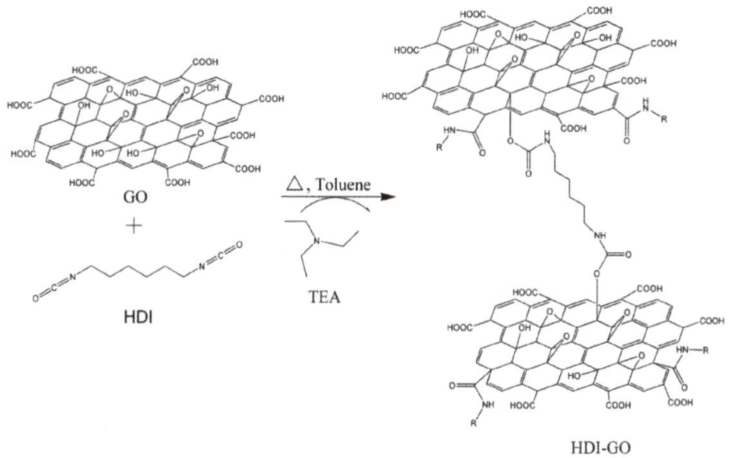

Scheme 1. Schematic representation of the synthesis procedure of hexamethylene diisocyanate (HDI)-functionalized graphene oxide (GO).

Different reactions conditions were tested in order to optimize the functionalization degree (FD), and detailed characterizations were conducted to confirm the successful functionalization of HDI on the GO surface. The resulting HDI-functionalized GO derivatives showed a more hydrophobic nature than pristine GO and could be suspended in polar aprotic solvents such as DMF, *N*-methylpyrrolidone (NMP) and DMSO, as well as in some low polar/non-polar solvents like tetrahydrofuran (THF), chloroform and toluene. In addition, the thermal stability was investigated to evaluate the effect of the functionalization on the thermal properties of the nanomaterials. Given that the FD of GO can be controlled by modifying the reaction conditions, it is feasible to synthesize GO sheets with partial HDI functionalization, which could be further subjected to a chemical reaction with other organic molecules or polymers via the residual hydroxyl groups. Further, depending on their FD, their solubility and dispersibility in different solvents can be tailored, which is interesting for the preparation of composites via mixing with polymers that have different solubilities. They could also be used in functional applications including electromagnetic wave absorption materials, electronic devices, field-effect transistors, memory devices, printable electronics, hydrogen storage as well as electrodes in lithium ion batteries, ionic conductors and supercapacitors [13,14]. Besides, due to their more hydrophobic features, they could enable the oil enrichment for the crude oil separation from seawater, and would be useful in coatings, biomedical devices, ultrasensitive sensors and biosensors.

2. Materials and Methods

2.1. Reagents

Natural graphite was obtained from Bay Carbon, Inc. (Bay City, MI, USA). H_2SO_4, $KMnO_4$, P_2O_5, $K_2S_2O_8$ and H_2O_2 (30 wt% in water) were purchased from Sigma-Aldrich (St. Louis, MO, USA) and used as received. HDI (>99%, $C_8H_{12}N_2O_2$, M_w = 168.196 g/mol) was acquired from Acros Organics (Pittsburgh, PA, USA). TEA (>98%, $N(CH_2CH_3)_3$, M_w = 101.193 g/mol) was obtained from Fluka Analytical (Munich, Germany). All the organic solvents were high performance liquid chromatography (HPLC) grade and were purchased from Scharlau S.L. (Barcelona, Spain). Toluene was dried and purified with a MBRAUN solvent purification system (Regensburg, Germany). Ultrapure water was obtained from a Millipore Elix 15824 Advantage 15 UV purification system (Bay City, MI, USA).

2.2. Synthesis of GO

GO was prepared using a modified Hummers' method from flake graphite [3,8]. Briefly, graphite powder, H_2SO_4, $K_2S_2O_8$, and P_2O_5, were heated at 80 °C for 5 h. After cooling, deionized water was added to the mixture and it was stirred overnight. The resulting product was then filtered, dried under air and oxidized again via addition of H_2SO_4, $KMnO_4$ and water in an ice-water bath. Following to dilution with water, excess $KMnO_4$ was decomposed by addition of 30 wt% H_2O_2 aqueous solution and then 5 wt% HCl aqueous solution. The product was filtered again and purified by repeating the following cycle: Centrifugation, removal of the supernatant liquid, addition of aqueous solution of H_2SO_4 (3 wt%)/H_2O_2 (0.5 wt%) and bath ultrasonication for 30 min at a power of 140 W. Then, it was washed several times with deionized water and finally vacuum freeze-dried before use.

2.3. Synthesis of HDI-Functionalized GO

The whole process was carried out under inert atmosphere of argon in order to avoid contamination during the functionalization reaction. In a typical experiment, GO powder (c.a. 250 mg) was weighed and loaded into a 100-mL round-bottom flask, followed by addition of dried toluene (25 mL) under Ar atmosphere. The suspension was then ultrasonicated in an ultrasonic bath for 2 h; in some experiments, the bath sonication was preceded by probe sonication cycles (5 min on/5 min off, 40% amplitude). The sonication conditions were chosen according to preliminary studies carried out in the group [22]. The GO dispersion was then transferred to a reactor equipped with mechanical agitator, thermometer and reflux condenser. Subsequently, TEA (c.a. 8.75 mL) and HDI (5 mL) were added dropwise via a dropping funnel. The mixture was heated to 60 °C and stirred at 350 rpm overnight under inert atmosphere. The resultant slurry reaction mixture was then poured into methylene chloride to coagulate the product, and finally filtered, washed thoroughly with methylene chloride and dried under vacuum to yield HDI-GO. A schematic representation of the synthesis procedure of functionalized HDI-GO is shown in Scheme 1.

The reaction conditions, namely reaction temperature, reaction time, GO/HDI/TEA ratio, tip/bath sonication cycles and solvent volume, were varied in order to determine their effect on the product yield. The conditions of each experiment and nomenclature of the different HDI-GO samples obtained herein are detailed in Table 1. Similar to pristine GO, the HDI-functionalized GO samples consisted in fine powders. With increasing functionalization degree, the color changed from dark black to metallic grey, indicative of loss of aromatic character.

Table 1. Nomenclature and reaction conditions for the synthesis of the different HDI-GO samples.

Entry	Sample	Reaction Time (h)	Reaction Temperature (°C)	GO/HDI/TEA Weight Ratio	Tip/Bath Sonication Time (min)	Solvent Volume (mL)
1	GO	-	-	-	-	-
2	HDI-GO 1	12	60	1/1/1	0/120	25
3	HDI-GO 2	12	60	0.5/1/1	0/120	25
4	HDI-GO 3	48	60	1/1/1	0/120	25
5	HDI-GO 4	12	90	1/1/1	0/120	25
6	HDI-GO 5	12	60	1/1/1	5/120	50
7	HDI-GO 6	12	60	1/1/1	5 + 5 + 5 */120	50

* 3 probe sonication cycles with 5 min of break between cycles.

2.4. Instrumentation

A Selecta 3001208 ultrasonic bath and a 24 kHz Hielscher UP400S ultrasonic tip processor (maximum power output of 400 W, Teltow, Germany) equipped with a titanium sonotrode with a diameter of 7 mm and length of 100 mm were used to prepare the GO and HDI-GO dispersions in the different solvents.

Elemental analysis was carried out with a LECO CHNS-932 elemental analyzer (Stockport, UK). Chromatograms were recorded on a gas chromatography/mass spectrometry (GC-MS, Santa Clara, CA, USA) turbo system (5975-7820A) model equipped with a HP-5MS capillary column (30 m × 0.25 mm × 0.25 µm), under the following conditions: Injector temperature 250 °C, detector 150 °C, oven temperature program: 50 °C ramp 5 °C/min until 100 °C, another ramp 10 °C/min until 230 °C (15 min).

Nuclear magnetic resonance (NMR) spectra were recorded using Varian Mercury 300 spectrometer (Santa Clara, CA, USA). Chemical shifts (δ) are reported in ppm and were measured relative by the internal referencing to the CDCl3 (1H).

Fourier-transformed infrared (FT-IR) spectra were recorded with a Perkin Elmer Frontier FTIR spectrophotometer (Waltham, MA, USA) equipped with an Attenuated Total Reflection (ATR) sampling accessory. Spectra were recorded at room temperature, in the wavenumber range of 500–4000 cm^{-1}, with an incident laser power of 1 mW and a minimum resolution of 4 cm^{-1}. Prior to the measurements, the powder samples were mixed and ground with KBr, and the mixtures were then pressed into a round transparent pellet in a pellet-forming dye.

Room temperature Raman spectra were obtained with a Renishaw Raman microscope (Gloucestershire, UK) equipped with a He-Ne laser (632.8 nm). The laser power at the sample was 1 mW. Three scans were recorded for each sample to reduce the signal-to-noise ratio. For comparative purposes, spectra were normalized to the G band.

Transmission electron microscopy (TEM) images were acquired with a Philips Tecnai 20 FEG (LaB6 filament) electron microscope (Philips Electron Optics, Holland) fitted with an EDAX detector, operating at 200 kV and with 0.3 nm point-to-point resolution. Samples were prepared by re-suspending in an ultrasonic bath about 2 mg of each sample powder in 5 mL of DMF. A drop of each suspension was cast on 200 mesh Lacey copper grids and dried under reduced pressure.

Water contact angle measurements were carried out at room temperature using a Ramé-Hart Model 500 Advanced Goniometer (Succasunna, NJ, USA). Prior to the measurements, the samples were dispersed in DMF and spin coated onto silane-functionalized glass substrates. A drop of Milli-Q water was deposited placed on the sample surface and the evolution of the droplet shape was recorded with a CCD video camera. DROPimage Advanced v2.4 analysis software was used to determine the contact angle. For each sample, the reported value is the average of the results obtained on three droplets.

The functionalization degree of HDI-GO and thermal stability of the samples was determined by thermogravimetric analysis (TGA) tests under a nitrogen atmosphere using a TA-Q500 thermobalance (Barcelona, Spain) coupled to a mass spectrometer, at a heating rate of 10 °C/min, from 100 to 600 °C.

3. Results and Discussion

3.1. Reaction Mechanism and Functionalization Degree

Organic isocyanates can react with both the edge carboxylic acid and surface hydroxyl or epoxide functional groups of GO through formation of amides or carbamate esters, respectively [23]. Isocyanate reactions are complicated owed to several reasons: (1) Competing side reactions, (2) reversible reaction steps, and (3) catalytic effect of reactants and products. Thus, it turns out to be quite difficult to establish a reaction mechanism and a kinetic model. The reactivity of isocyanate vs. different functional groups follows the order: Primary amine > secondary amine > hydroxyl > acid > anhydride > epoxide [24]. The reaction between an alcohol and an isocyanate group to form a carbamate ester is moderate, hence is usually catalyzed by bases, mainly tertiary amines like TEA [25]. The catalytic activity of amines is attributed to the presence of a lone pair on the nitrogen atom. Several mechanisms regarding the catalytic activity of amines can be found in the literature [25–27], the most important ones being the formation of an isocyanate-amine complex or an active hydrogen-amine complex (Scheme 2).

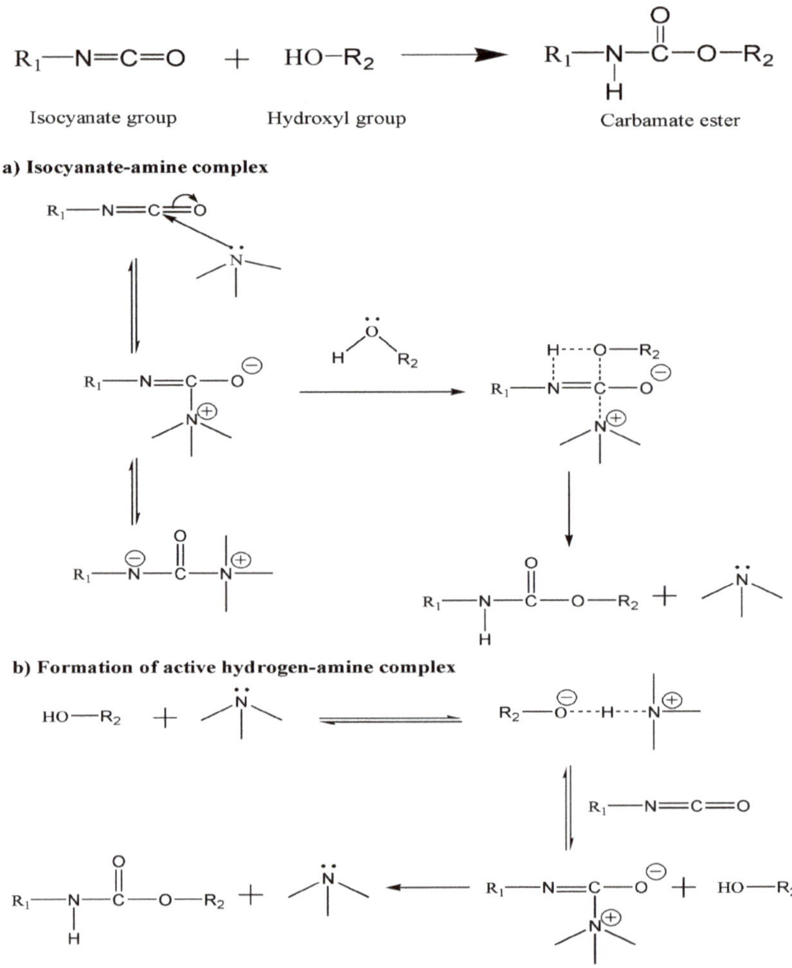

Scheme 2. Mechanism of the reaction between an isocyanate and a hydroxyl group to form a carbamate ester catalyzed by ternary amines: (**a**) Formation of an isocyanate-amine complex; (**b**) formation of active hydrogen-amine complex.

The first mechanism (Scheme 2a), proposed by Baker and Holdsworth [27] involves the reversible nucleophilic attack on the carbon atom of the isocyanate by the amine to form a complex in which the nitrogen of the NCO group is activated, making easier the reaction with the lone electrons of the oxygen of the alcohol. Then, the complex is decomposed, forming the carbamate ester and regenerating the base. Another mechanism proposed by Farkas and Strohm [25] involves the activation of the alcohol via hydrogen bonding with the amine, resulting in the formation of a complex that subsequently reacts with the isocyanate group (Scheme 2b).

Isocyanates can also react with carboxylic acids via an amidation reaction, albeit the yield in conventional organic solvents is typically low [28]. Besides, the side reaction of isocyanates with water results in amine formation and CO_2 gas release. The reaction is exothermic and is also catalyzed by tertiary amines. It has been demonstrated that the isocyanate reactivity with water and primary hydroxyl group is comparable, and more than double that with carboxylic acids [24]. Isocyanates also react with amines to give ureas. On the other hand, the epoxide functionality is the least reactive; the reaction between isocyanate and epoxide to form carbamate ester linkages is unlikely and only takes place in the presence of special catalysts [29]. From all the above discussion, and taking into account the IR spectra of the different HDI-GO samples (Figure 1), which show the appearance of an intense absorption peak at ~1710 cm^{-1} ascribed to the C=O stretching vibration of carbamate esters [20], as will be discussed in detail in the following section, it can be concluded that the formation of carbamate esters via reaction of HDI with surface OH groups of GO is the most likely.

Given that HDI possesses two reactive isocyanate groups, it can act as a crosslinking agent, joining adjacent GO layers (Scheme 1). A very high level of crosslinking is not desirable, since it can make the exfoliation of the functionalized GO into individual sheets difficult, thus hindering subsequent reactions with other organic molecules or polymers via the remaining oxygen groups. The crosslinking between the GO sheets that are randomly covalently bonded by HDI to each other combined with the higher hydrophobic character of the HDI-GO compared to GO will influence their physical properties (i.e., solubility, polarity, thermal stability, etc.), as will be discussed in the following sections.

The results obtained from elemental analysis of neat GO and the different functionalized samples are collected in Table 2. Assuming that the formation of carbamate esters through reaction of HDI with the OH surface functional groups of GO is the unique reaction path, the nitrogen-to-carbon atomic ratio can be used to roughly estimate the functionalization degree (FD). In particular, FD can be expressed as moles of carbamate ester unit incorporated per mol of carbon atoms of GO, and the calculated values are tabulated in Table 2. It should be noted that these values correspond to a lower bound of functionalization; in the case of amide formation, which seems less feasible according to the IR spectra (Figure 1) since there is no peak at around 1680–1650 cm^{-1} related to the amide I band, the FG would be higher due to a loss of carbon (as carbon dioxide) from the isocyanate reagent.

Table 2. Elemental analysis data, functionalization degree (FD) and water contact angle (CA) for neat GO and the different HDI-GO samples.

Entry	Sample	C (%)	O (%)	H (%)	N (%)	S (%)	FD * (%)	CA (°)
1	GO	41.93	51.96	3.44	0	2.67	0	49.5
2	HDI-GO 1	53.08	35.70	4.22	6.02	0.98	12.28	75.6
3	HDI-GO 2	47.38	44.75	3.83	2.49	1.55	5.08	58.7
4	HDI-GO 3	50.36	40.16	4.01	4.46	1.01	9.10	68.6
5	HDI-GO 4	45.98	46.97	3.67	1.53	1.85	3.12	54.3
6	HDI-GO 5	55.49	31.07	4.50	8.43	0.51	17.20	89.8
7	HDI-GO 6	56.04	30.09	4.55	8.88	0.44	18.13	93.5

* moles of carbamate ester unit incorporated per mol of carbon atoms of GO.

To understand the data obtained, the factors that influence the chemical kinetics of a reaction, namely reactant concentrations, temperature, physical states and surface areas of reactants, solvent and catalyst concentration and properties, and so forth, have to be taken into account. The kinetics and mechanism of tertiary-amine catalyzed reaction between isocyanates and primary alcohols have been investigated by several authors [27,30–33], and it is commonly accepted that the reaction proceeds by competitive consecutive second order reactions through a carbamate ester-isocyanate intermediate; besides, the reaction rate was found to be directly proportional to the concentration of each reactant and the amine catalyst [32].

Given that it is not possible to determine the molarity of GO because it is a non-stoichiometric compound and it has no specific molecular weight, it was not feasible to reproduce the concentration ratios of reactants and catalyst previously reported for the reactions of isocyanates with alcohols.

Therefore, the first trial (HDI-GO 1) was carried out with GO:HDI:TEA weight ratios of 1:1:1, leading to a FD of around 12% (Table 2, entry 2). Surprisingly, the increase in the amount of HDI reactant and TEA catalyst (HDI-GO 2, with GO:HDI:TEA weight ratios of 0.5:1:1) results in a drop in FD (Table 2, entry 3). The reaction extent is likely limited by the number of hydroxyl groups on the GO surface available for reaction. Hence, the increase in the HDI concentration and TEA does not lead to a higher FD. It is well known that the molecules of TEA do not associate with each other albeit form hydrogen bonds with alcohols to some extent. Since TEA acts as a catalyst, when its concentration is increased, the excess of molecules likely interact with the OH surface groups of GO via hydrogen bonding, hence competing with or hindering the reaction between the hydroxyl moieties and HDI, consequently the reaction yield does not increase or even decreases. This is consistent with the observations reported earlier by other authors [31,33], who found that the reactivity of the OH groups depended on whether they were free or hydrogen bonded, because the activation energies of hydrogen bonded alcohols for carbamate ester formation were considerably higher than those of free alcohols. Further, Sato [31] found a decrease in the reaction rate with increasing the amount of reactants. It was suggested that when the concentration of isocyanate was doubled, it spontaneously reacted with the moisture contained in the solvent or present in the environment and with the carbamate to produce a small amount of the corresponding urea, thus the reaction yield was reduced. In fact, alkyl-alkyl urea has been reported to be a major byproduct in the reactions of aliphatic isocyanates with alcohols [34]. However, in this study the synthesis of HDI-GO was carried out under an Ar atmosphere and with dried toluene, hence the formation of urea byproduct, in particular N,N'-dihexylurea, is not likely. Further, under some circumstances in the presence of amine catalysts, isocyanates can form dimers, trimers or polymerize [35] by linear polymerization and condensation with elimination of carbon dioxide to form uretidinone, isocyanurate, and carbodiimide. In particular, traces of N,N'-dihexylcarbodiimide could be formed by decomposition of HDI dimmers [35].

To corroborate the absence of urea and carbodiimide byproducts, the toluene removed during the filtration stage was analyzed by gas chromatography/mass spectrometry and RMN techniques. The chromatogram showed only one compound and the mass spectra revealed that this peak corresponded to the initial product. The ^{1}H NMR spectrum exhibited just three peaks related to protons of $OCNCH_2$ [3.31 ppm, t], $OCNCH_2CH_2$ [1.70–1.55 ppm, m] and $OCNCH_2CH_2CH_2$ [1.49–1.31 ppm, m]; these chemical shifts are in perfect agreement with those of the initial HDI, which confirms the lack of byproducts in the synthesis of HDI-GO.

Regarding the reaction carried out for 48 h (HDI-GO 3), a slight fall in FD is found (Table 2, entry 4) compared to the initial trial. Lu et al. [24] reported that the conversion yield of the reaction between an isocyanate and an alcohol crosses over at longer reaction times, suggesting partial reversibility and equilibrium. Similarly, Sato [31] found that at high concentrations of both reactants, the reaction rate decreased rapidly with increasing reaction time. These facts could explain the small drop in FD at very long reaction times.

On the other hand, the rise in the reaction temperature (HDI-GO 4) also resulted in a lower FD (Table 2, entry 5), which was only about 3%. This is consistent with the results reported previously for the reaction of n-alcohols with phenyl isocyanate, where the rate of the reaction vs. $1/T$ followed a straight line [30]. This behaviour could be rationalized considering that the reaction between an alcohol and an isocyanate is exothermic, hence when the temperature is increased, the yield of products is decreased. Besides, it has been reported that at elevated temperatures the reaction becomes rapidly reversible [36], giving isocyanate and alcohol. This is in agreement with the IR spectrum of HDI-GO 4, which shows a very intense O–H stretching peak at 3430 cm^{-1} and N=C=O stretching of the isocyanate groups at 2260 cm^{-1}. These peaks gradually drop in intensity with increasing FD, indicative of a lower extent of the reverse reaction.

Another factor that could influence the reaction yield is the solvent volume. Given that the functionalization reaction should take place on the surface groups of GO, the HDI and TEA have to diffuse from the bulk solution to the GO surface, and the diffusion is typically the slowest stage that controls the overall reaction rate. The solvent facilitates the diffusion of the reactant and catalyst towards the nanomaterial surface; if the solvent volume is too low, these cannot effectively diffuse, and the reaction becomes saturated. Therefore, low solvent volumes likely limit the reaction yield. Further, the solvent molecules arrange around the reactants forming solvent cages. This "cage effect" hinders the separation of the reactant molecules and easily absorbs the excess energy of products, thus favouring product formation [37]. The increased number of collisions caused by a stronger solvent cage effect would increase the reaction rate.

To test this hypothesis, further experiments were carried out with double volume of solvent (HDI-GO 5 and HDI-GO 6). Besides, to enlarge the surface area of GO available for the reaction, it was subjected to sonication with an ultrasonic probe, a process that effectively induces the exfoliation of the nanomaterial into thinner sheets. However, this procedure generates defects in the GO flakes: Point defects on the basal plane and edge defects, which may have detrimental effects on the nanomaterial properties, and the number of defects rises with increasing ultrasonication time [38]. Therefore, an optimum balance between level of exfoliation and defect density has to be attained. For such purpose, the probe sonication was combined with bath ultrasonication, which is less aggressive. The probe sonication conditions (5 min at 40% amplitude) were chosen according to previous studies carried out in the group [22]. By only applying one probe sonication cycle (HDI-GO 5), a FD of about 17% was attained (Table 2, entry 6). Another experiment was performed by applying three probe sonication cycles with a 5 min break between cycles, leading to a FD close to 18% (Table 2, entry 7). It is therefore confirmed that the longer the sonication with the probe, the better the level of GO exfoliation, the larger the surface area on which collisions can occur, hence the most effective the reaction is. Nonetheless, the differences in the extent of the reaction between HDI-GO 5 and HDI-GO 6 are relatively small, hinting that a saturation level has been attained and the FD has been leveled off, hence these were taken as the optimum ones.

3.2. FT-IR Study

The chemical changes that occurred upon treatment of GO with HDI were monitored by FT-IR spectroscopy, since both GO and the derivatives display characteristic IR spectra (Figure 1). As mentioned earlier, GO contains epoxide and hydroxyl functional groups on both sides of its basal plane and carboxyl moieties at the edge sites. The most characteristic features in the IR spectrum of GO are the strong broad band centered at ~3430 cm^{-1} corresponding to the O–H stretching vibrations, the peaks at around 2925 and 2845 cm^{-1} attributed to sp^2 and sp^3 C–H stretching bands produced at defects sites of the graphene network, the peak at ~1730 cm^{-1} arising from the C=O stretching of the carboxylic acid groups, the band at 1620 cm^{-1} assigned to the aromatic C–C stretching, that at ~1400 cm^{-1} corresponding to the O–H deformation [39] and the C–OH stretching at 1260 cm^{-1} [3].

Upon treatment with HDI, the intensity of the O–H stretching band was reduced, decreasing gradually with increasing FD, and also shifted towards lower wavenumbers, due to the overlapping with the N–H stretching vibrations of the carbamate groups. Further, the peaks at 2845 and 2925 cm^{-1} originating from symmetrical and asymmetrical stretching vibrations of –CH$_2$– became more intense upon raising FD, due to an increased number of methylene chains arising from the HDI. A new band appeared in the samples with low FD at 2280 cm^{-1} ascribed to unreacted N=C=O group [40], suggesting the absorption/intercalation of the organic isocyanate between the GO flakes. However, this band can hardly be detected in the HDI-GO 5 and HDI-GO 6 samples, corroborating that the –N=C=O groups of HDI reacted completely with the hydroxyl groups of GO. Further, the C=O stretching appearing at 1730 cm^{-1} in pristine GO is hidden by a new peak at ~1710 cm^{-1} ascribed to the C=O stretching of the carbamate ester groups [20]. This peak becomes more intense and shifts gradually towards lower wavenumbers with increasing FD.

Besides, new intense bands can be observed at ~1648 and 1580 cm^{-1}; the first can be assigned to the coupling of the C=O stretching with in-phase N–H bending and the second to the coupling of the N–H bending with the C–N stretching vibration [40]. These bands could originate from either amide or carbamate ester groups, although in the first case they typically appear at 1660 and 1550 cm^{-1}, whilst for carbamate esters the bands are closer together due to the stronger π-π interaction between the carbonyl group and the nitrogen lone pair electrons, and the amide II band appears at higher frequency [41]. These two peaks also turn out to be stronger and move to lower wavenumbers with increasing FD, which could be indicative of increased H-bonding interactions between carbamate esters closely located [42]. Other bands become visible at around 1110 cm^{-1}, likely related to –C(=O)–O and C–N stretching vibrations of the carbamate groups, at 885 cm^{-1} attributed to C–H out-of-plane bending vibrations of substituted aromatic rings and at around 720 cm^{-1}, ascribed to the rocking of the methylene groups of HDI. On the basis of all the aforementioned observations, it can be concluded that GO was successfully functionalized with the organic HDI reactant and that the functionalization route via carbamate ester formation predominates.

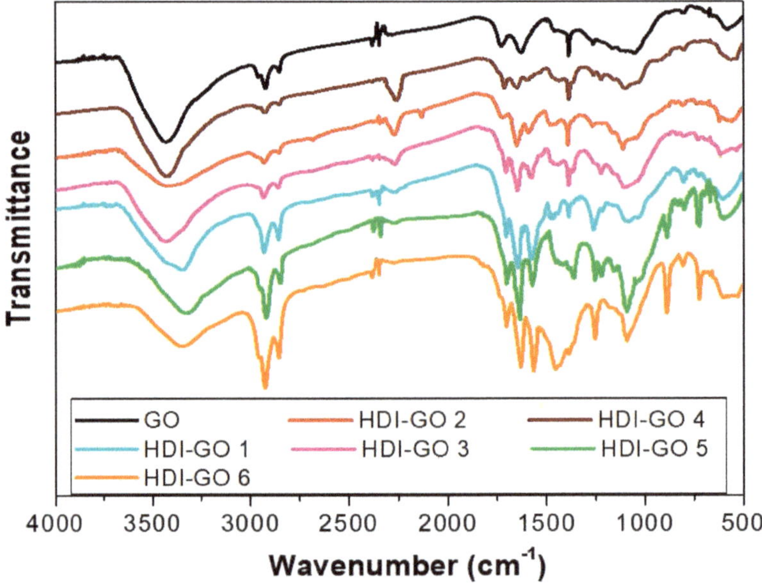

Figure 1. Fourier-transformed infrared (FT-IR) spectra of raw GO and the different HDI-GO samples.

3.3. Water Contact Angle Measurements

Water contact angle (CA) measurements were performed to evaluate the hydrophobic or hydrophilic character of the synthesized HDI-GO. In general, surfaces with contact angles ≥90° are regarded as hydrophobic, or in other words water-repellent. Figure 2 compares the contact angles for GO (a), HDI-GO 2 (b), HDI-GO 1 (c) and HDI-GO 6 (d), and the CA values for all the HDI-GO samples synthesized herein are summarized in Table 2. Raw GO is highly hydrophilic (CA ~50°) due to its large number of surface oxygen-containing groups that lead to an immediate absorption of water and swelling of the sample. A gradual increase in CA is found upon increasing FD, indicating lower level of hydrophilicity (wettability), ascribed to the decrease in the number of surface hydroxyl groups due to their reaction with HDI and the higher degree of crosslinking between the GO flakes that reduces the swelling capacity [43]. The HDI-GO samples with FD ≤12% show CA values in the range of 54–75°, hence can be still regarded as hydrophilic like the parent GO. It is interesting to note that the most favorable CA values for protein adsorption and cell adhesion have been reported to be in the range of 40–70° [44], which suggests that the derivatives with low functionalization degree could be suitable for biomedical applications. On the other hand, the HDI-GO 5 and HDI-GO 6 samples have a CA ≥90°, hence are hydrophobic in nature. Overall, it is found that the functionalization reaction with HDI leads to an increased surface hydrophobicity and reduces the hydration of the remaining hydroxyl groups.

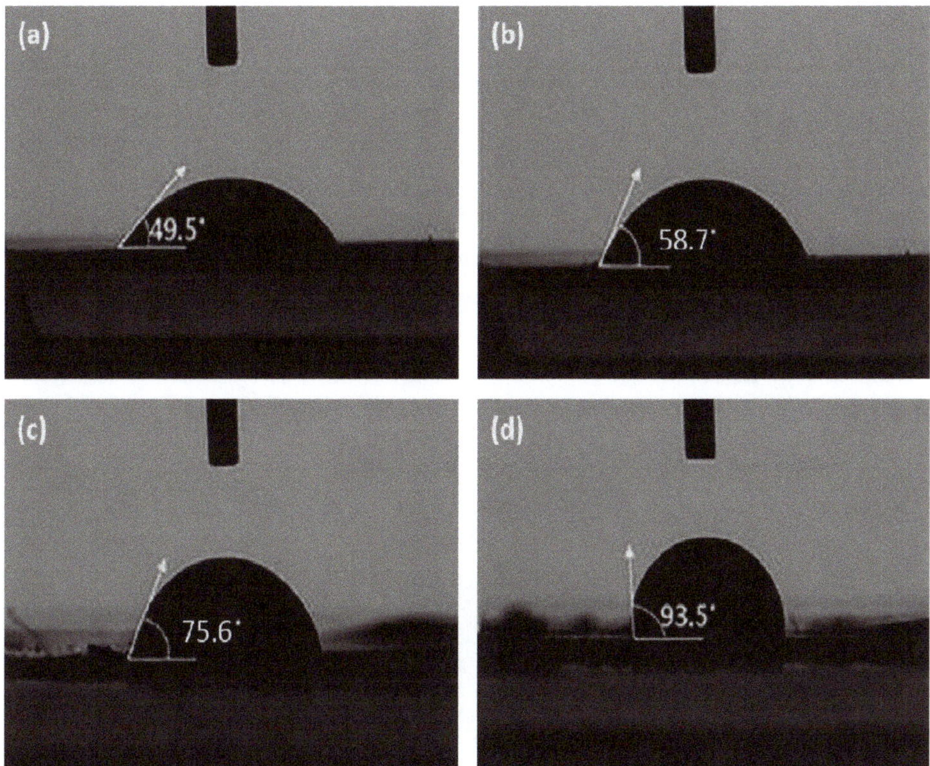

Figure 2. Water contact angle measurements of GO (**a**); HDI-GO 2 (**b**); HDI-GO 1 (**c**); and HDI-GO 6 (**d**).

3.4. Solubility Tests

As detailed in the introduction, the stability of HDI-GO in a range of solvents is a critical point for the preparation of dispersions to be applied in various fields. In this context, the solubility/dispersibility of the functionalized samples was investigated in water and organic solvents of different nature, and directly compared with that of pristine GO. The following organic solvents were tested: Acetone, methanol, ethanol, 2-propanol, THF, DMF, NMP, DMSO, chloroform, toluene, n-hexane and n-pentane. For solubility tests, a small quantity of GO or HDI-GO (~0.5 mg) was added to a given volume of solvent (~1 mL) in a test tube and gently stirred with a glass stirring rod for a few min. To check the dispersibility, the samples were sonicated in an ultrasound bath for 10 min. The results obtained from the solubility tests are summarized in Table 3, and representative photographs of the GO and HDI-GO 6 dispersions in water, DMF, NMP, toluene and n-hexane are shown in Figure 3.

Pristine GO is completely soluble in water, propanol, THF and NMP (Table 3, entry 1), in agreement with the results reported previously [45]. However, despite its high oxygen content (~52%, Table 1), it is only partially soluble in polar protic solvents, such as methanol or ethanol (Table 3, entry 8). This indicates that the solvent polarity is not the only issue controlling the solubility. Other factors such as the solvent dipole moment, its surface tension and the Hansen solubility parameters play a key role on the solubility behaviour [46]. Thus, the best solvents exhibit high dipole moment values (i.e., 1.85, 1.66 and 3.75 for water, propanol and NMP [46]) and a surface tension close to the reported surface energy of GO (~62 mN/m [47]). It is worth mentioning that GO showed low solubility but stable dispersibility in non-polar solvents, like toluene or chloroform, while was insoluble in hexane or pentane.

Table 3. Solubility of GO and the different HDI-GO samples in different solvents.

Entry	Sample	Water	DMF	NMP	DMSO	CH_3Cl	Toluene	THF
1	GO	S	PS/SS	S	PS/SS	PS/SS	PS/SS	S/PS
2	HDI-GO 1	I	PS	PS	PS	I	I	SS
3	HDI-GO 2	PS	PS/SS	S	PS	SS	SS	PS
4	HDI-GO 3	SS	PS	PS	PS	I	I	SS
5	HDI-GO 4	PS	PS/SS	S	PS/SS	SS	SS	PS
6	HDI-GO 5	I	PS	PS	PS	I	I	I
7	HDI-GO 6	I	PS	PS	PS	I	I	I

Entry	Sample	Acetone	Pentane	Hexane	Methanol	Ethanol	Propanol	FD
8	GO	I	I	I	SS	PS	S/PS	0
9	HDI-GO 1	PS	I	I	I	I	SS	12.28
10	HDI-GO 2	I	I	I	PS	PS	PS	5.08
11	HDI-GO 3	SS	I	I	SS	SS	PS	9.10
12	HDI-GO 4	I	I	I	PS	PS	S/PS	3.12
13	HDI-GO 5	PS	I	I	I	I	I	17.20
14	HDI-GO 6	PS	I	I	I	I	I	18.13

S: Soluble; PS: Partially soluble; SS: Slightly soluble; I: Insoluble.

Figure 3. Photographs of the dispersions of GO (**top**) and HDI-GO 6 (**bottom**) in different solvents: (**a**) Water; (**b**) DMF; (**c**) NMP; (**d**) toluene; (**e**) n-hexane.

In contrast to neat GO, the HDI-GO samples were hardly soluble in water and polar protic solvents, and the solubility decreased with increasing FD. However, upon a short ultrasonication treatment, they swelled and formed colloidal dispersions in polar aprotic solvents such as DMF, NMP and DMSO, and the dispersibility improved with increasing FD. The HDI-GO samples with higher FD showed better dispersibility in DMF and DMSO than the parent GO, and were stable for a few weeks. Owed to their higher hydrophobic character, they could also be more easily suspended in low polar solvents like THF or non-polar ones such as chloroform and toluene (Figure 3d). The reduction of hydrogen bonds triggered by the functionalization with HDI renders the GO surfaces more hydrophobic, and the strength of hydrogen bonding between the HDI-GO layers is weakened, thus allowing for improved dispersion in organic solvents. Overall, DMF and DMSO were chosen as optimum solvents to prepare HDI-GO dispersions for further reaction with conductive polymers.

3.5. TEM Analysis

The surface morphology of neat GO and the different functionalized samples was examined by TEM, and typical images at different magnifications of GO, HDI-GO 2, HDI-GO 5 and HDI-GO 6 are compared in Figure 4.

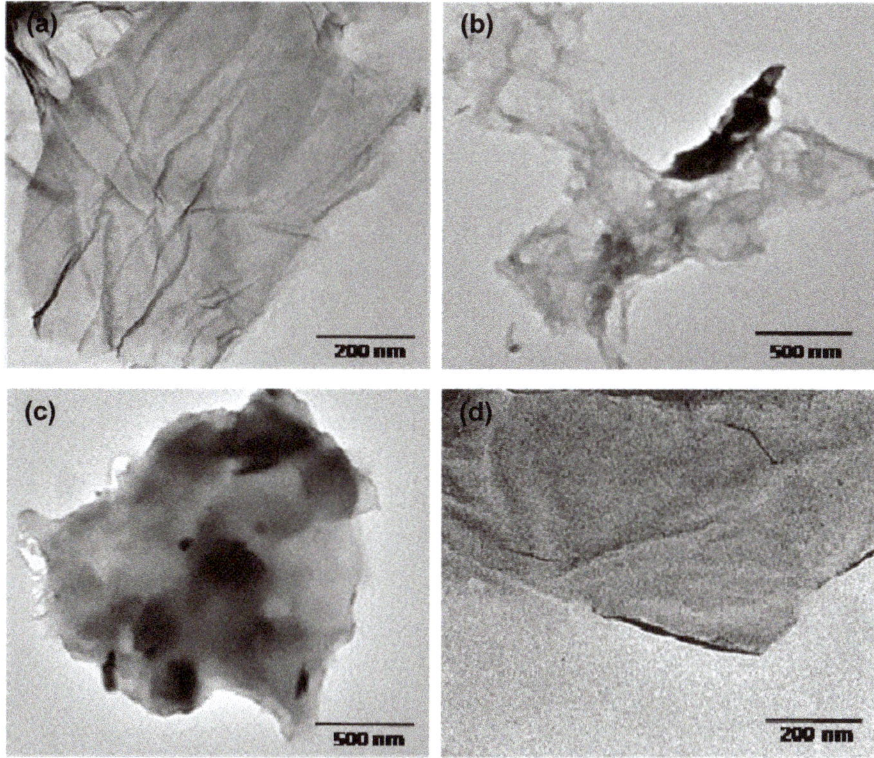

Figure 4. Typical transmission electron microscopy (TEM) images at different magnifications of GO (**a**); HDI-GO 2 (**b**); HDI-GO 5 (**c**); and HDI-GO 6 (**d**).

The thickness of a single monolayer graphene nanosheet has been reported to be about 0.34 nm [2], and a GO sheet is expected to be much thicker due to the presence of oxygen-containing functional groups attached on both sides of the flakes. Pristine GO displays a transparent and pleated sheet structure (Figure 4a), with wrinkled and highly flexible flakes of a thickness ≤10 nm. The sample appears essentially homogeneous and uniform, comprising a few oxidized graphene layers.

Conversely, the HDI-GO samples seem more heterogeneous, and present a relatively fuzzy morphology due to the grafting of the HDI chains onto the nanomaterial surface. Apparently, the darker areas in the images correspond to the alkyl chains of the organic diisocyanate, since the number of dark areas increases with increasing FD. This is consistent with the results reported previously by other authors who grafted alkyl segments onto G via reaction with surfactant molecules [48,49] or polymeric chains as described elsewhere [50]. According to those studies, the grafted alkyl chains are extended in solution due to their solubility in the solvent; nonetheless, upon drying the alkyl segments collapse onto the surface of the GO sheets, forming nanosized domains that likely correspond to the observed dark spots.

In the samples with higher FD, a morphology of fluffy and smooth surfaces in a stack state can be observed (Figure 4c), ascribed to the alkyl chains irregularly arranged on the GO surface, particularly at the wrinkles and edges that are more reactive owed to the lattice defects. The surface foldings cannot be observed in these functionalized samples, likely due to the wrapping effect of the organic moieties attached to the GO which, thus, enshroud these wrinkles. Besides, the surface of the HDI-GO samples is coarser than that of GO, with some visible defects, albeit retaining the structural integrity of the graphene framework, and the flake thickness increases with increasing FD.

It should be noted that GO is non-homogeneously covered by the HDI chains in the derivatives with low FD (Figure 4b), whereas the HDI-GO 6 shows a much more uniform coating (Figure 4d), in agreement with its higher extent of functionalization. The GO surface appears fully covered by a spotted dark pattern, suggesting that a complete coverage has been attained; hence it would be difficult to obtain a higher FD.

3.6. Raman Spectra

Raman spectroscopy was carried out to verify the structural changes in GO upon reaction with HDI, and the spectra of the different samples are compared in Figure 5. The most important features in the Raman spectrum of graphene materials are the disorder induced D band at 1345 cm^{-1} that indicates the level of structural disorder and the tangential G band at ~1595 cm^{-1} arising from in-plane displacements in the graphene sheets.

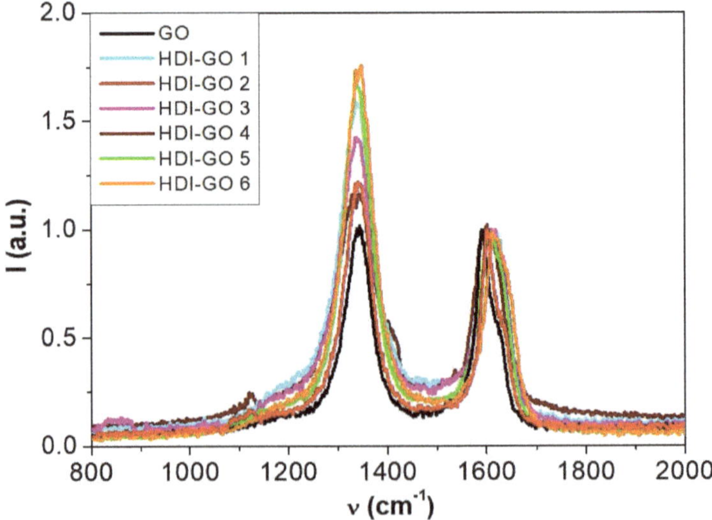

Figure 5. Raman spectra of GO and the different HDI-GO samples.

The D to G band intensity ratio (I_D/I_G) is widely used to obtain quantitative information about the amount of defects in graphene materials: The higher the ratio, the higher the disorder [51]. The calculated I_D/I_G data for raw GO and the HDI-GO samples are summarized in Table 4, along with the position of the G band. A steady increase in this ratio is observed with increasing FD, from about 1 for GO (Table 4, entry 1) to 1.75 for the derivative with the highest FD (Table 4, entry 7), which demonstrates a decrease in the structural order, that is, in the size of the in-plane sp^2 domains, owed to the covalent grafting of the HDI chains onto the GO surface. An analogous trend of I_D/I_G increase has been reported upon covalent functionalization of GO with other organic molecules such as 4-aminobenzenesulfonic acid [52] or polymers like poly(N-isopropylacrylamide) (PNIPAM) [53], hinting that the functionalization process results in graphitic domains that are smaller than those of GO.

Table 4. I_D/I_G ratio and position of the G band obtained from the Raman spectra as well as thermogravimetric analysis (TGA) data of the different samples.

Entry	Sample	T_i (°C)	T_{10} (°C)	T_{max} I/II (°C)	FD (%)	I_D/I_G	G (cm^{-1})
1	GO	124.1	180.6	235.0	-	1.01	1595
2	HDI-GO 1	165.4	230.7	250.5/396.2	21.9	1.57	1611
3	HDI-GO 2	141.3	199.5	248.6/374.6	7.6	1.22	1603
4	HDI-GO 3	148.6	213.6	245.6/390.6	16.1	1.40	1609
5	HDI-GO 4	130.2	189.2	241.9/385.5	4.5	1.15	1600
6	HDI-GO 5	174.5	242.8	250.9/404.2	29.2	1.66	1613
7	HDI-GO 6	186.3	288.3	249.7/406.6	30.8	1.75	1617

T_i: Initial degradation temperature at 2% weight loss; T_{10}: Temperature of 10% of weight loss; T_{max}: Temperature of maximum rate of weight loss. The subscripts I and II refer to the first and second degradation stages. FD: Functionalization degree obtained from TGA thermograms.

It can also be observed from the comparison of the Raman spectra that upon increasing FD the G-band of GO is broadened and upshifted to higher wavenumbers, by up to 22 cm^{-1} for HDI-GO 6 (Table 4, entry 7) compared to pristine GO (Table 4, entry 1). The change in the position of the G band is typically associated with a modification of the electronic structure of the carbon nanomaterial [54], being shifted to higher frequencies in the presence of electron-acceptor groups. Albeit the electron distribution of the isocyanate group makes it either an electron donor or electron acceptor, in the reaction performed herein it seems to act as an electron-acceptor (Scheme 2), thus causing a blue shift of the G band of GO. The blue shift could also be indicative of the damage of the hexagonal network of carbon atoms, thus a marker of the defect density, since the position of this G band has been reported to shift almost linearly with increasing the concentration of defects in the graphene layer [55]. Besides, upshifts in the G band of graphene-based materials have been reported in the presence of polymers [22] and surfactants [51], ascribed to intense polymer (or surfactant)-graphene interactions. In addition, the G mode peak position depends on the number of GO layers: It shifts to higher frequency as the number of GO layers decreases [55]. Therefore, the steady upshift found here upon increasing FD suggests a reduction in the degree of exfoliation of the GO flakes due to the constraints imposed by the crosslinks between the nanosheets, and consequently, could be indicative of a transition from few-layered to multi-layered GO. This is in agreement with TEM analysis (Figure 4), which revealed a raise in layer thickness with increasing FD.

3.7. Thermal Stability

Figure 6 compares the TGA thermograms under a nitrogen atmosphere of neat GO and the different HDI-GO samples. The pristine nanomaterial shows a one-step degradation process that starts (T_i) at 124 °C and exhibits the maximum rate of weight loss (T_{max}) at ~235 °C (Table 4, entry 1), with about 37% weight loss below 250 °C. This major weight loss is ascribed to the decomposition of its surface oxygen-functional groups, namely epoxide, hydroxyl and carboxylic acid, as revealed by the IR spectrum. Besides, a small gradual mass loss is found above 250 °C, attributed to the removal of additional functional groups [3]. In contrast, the HDI-GO samples display two major stages. The first stage is analogous to that found for neat GO, and should be due to the pyrolysis of the residual oxygen functional groups. The weight loss of this step, that is, the amount of surface oxygen moieties in the functionalized GO, gradually decreases with increasing FD, which can be explained considering that the higher the FD of the HDI-GO samples, the greater the number of oxygenated functional groups of GO that have reacted with the isocyanate groups of HDI, hence the lower the number of residual oxygenated groups on the GO surface that decompose during this stage. Therefore, this decreasing trend in weight loss with increasing FD corroborates the larger extent of the functionalization reaction, in agreement with the results from FT-IR analysis (Figure 1).

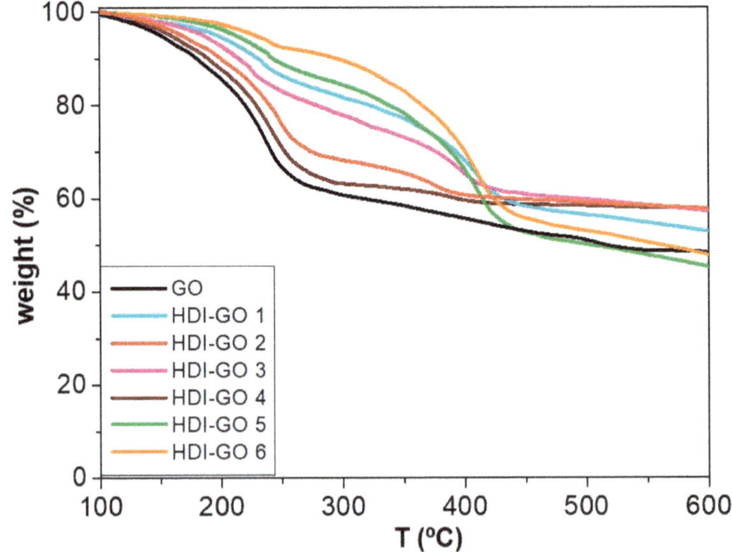

Figure 6. TGA curves under a nitrogen atmosphere of GO and the synthesized HDI-GO.

The second stage corresponds to the thermal decomposition of covalently grafted HDI on the surface of GO, as previously reported for other alkyl-modified GO derivatives [50], and its mass loss rises with increasing the degree of grafting. From this stage the extent of covalent functionalization or FD of each HDI-GO sample was determined, and the results are collected in Table 4. The values obtained follow exactly the same trend to those derived from elemental analysis, albeit the data from TGA thermograms are systematically higher (i.e., about 1.5 fold larger), which can be explained considering the different meaning of the FD provided by each technique.

Besides, the characteristic degradation temperatures, namely T_i, T_{max} and the temperature of 10% weight loss (T_{10}), grow steadily as the extent of functionalization increases (Table 4), which could be due to the higher degree of crosslinking between the nanosheets, as previously reported for chitosan/GO crosslinked composites [56]. Increasing the amount of crosslinking causes less of the sample to decompose in the first step. Thus, T_i increased by about 80 °C from neat GO (Table 4, entry 1) to the HDI-GO 6 sample (Table 4, entry 7); since HDI has two isocyanate groups, it can act as a crosslinking agent between the GO sheets, which is reflected in higher decomposition temperature. The TGA results confirm that the covalent attachment of HDI greatly improves the thermal stability of GO, which could be interesting for a variety of applications including long-term electronics or electrothermal devices.

4. Conclusions

HDI-GO derivatives with different functionalization degrees have been synthesized following a two-step approach: Firstly, GO was prepared using a modified Hummers´ method from graphite, and secondly GO was treated with HDI in the presence of a TEA catalyst to yield the modified nanomaterial. The FT-IR and Raman spectra corroborated the success of the reaction and that the functionalization route via carbamate ester formation predominated. The increase in the amount of HDI reactant and TEA catalyst, as well as the increase in the reaction time or temperature resulted in a drop in the extent of functionalization, whilst the combination of probe and bath sonication with higher solvent volumes led to the highest functionalization degrees. The HDI-GO displayed a more hydrophobic character than raw GO and could be dispersed in polar aprotic solvents such as DMF, NMP and DMSO as well as in some low polar/non-polar solvents like THF, chloroform and toluene. Further, the dispersibility in these solvents improved as the functionalization degree increased. The surface morphology of the samples was analyzed by TEM, and the images showed a raise in layer thickness with increasing extent of functionalization. TGA study revealed that the covalent grafting of HDI enhances the thermal stability of GO, ascribed to the crosslinking between neighbouring sheets. This thermal stability improvement is highly attractive for applications such as long-term electronics and electrothermal devices. Future work will focus on the reaction of HDI-GO with other organic molecules or polymers via the remaining oxygen groups, in order to develop high-performance GO-based nanocomposites.

Author Contributions: J.A.L.-S. performed the experiments and analyzed part of the data; A.M.D.-P. designed the experiments, supervised the work and wrote the paper; G.M. and C.G.-A. collaborated in the development of the experiments, in the analysis of the experimental data and the discussion of the results.

Funding: This research received no external funding.

Acknowledgments: Financial support from Fundación Iberdrola España via a Research Grant in Energy and the Environment 2017 is gratefully acknowledged. Dr. A. M. Diez-Pascual would like to thank the Ministerio de Economía, Industria y Competitividad for a "Ramón y Cajal" postdoctoral fellowship.

Conflicts of Interest: The authors declare no conflict of interest.

References

1. Geim, A.K.; Novoselov, K.S. The rise of graphene. *Nat. Mater.* **2007**, *6*, 183–191. [CrossRef] [PubMed]
2. Novoselov, K.S.; Geim, A.K.; Morozov, S.V.; Jiang, D.; Zhang, Y.; Dubonos, S.V.; Grigorieva, I.V.; Firsov, A.A. Electric Field Effect in Atomically Thin Carbon Films. *Science* **2004**, *306*, 666–669. [CrossRef] [PubMed]
3. Diez-Pascual, A.M.; Diez-Vicente, A.L. Poly(propylene fumarate)/polyethylene glycol-modified graphene oxide nanocomposites for tissue engineering. *ACS Appl. Mater. Interface* **2016**, *8*, 7902–17914. [CrossRef] [PubMed]
4. Díez-Pascual, A.M.; Gómez-Fatou, M.A.; Ania, F.; Flores, A. Nanoindentation in Polymer Nanocomposites. *Prog. Mater. Sci.* **2015**, *67*, 1–94. [CrossRef]
5. Salavagione, H.; Díez-Pascual, A.M.; Lázaro, E.; Vera, S.; Gomez-Fatou, M. Chemical sensors based on polymer composites with carbon nanotubes and G. The role of the polymer. *J. Mater. Chem.* **2014**, *2*, 14289–14328. [CrossRef]
6. Díez-Pascual, A.M.; Luceño Sánchez, J.A.; Peña Capilla, R.; García Díaz, P. Recent advances in graphene/polymer nanocomposites for applications in polymer solar cells. *Polymers* **2018**, *10*, 217. [CrossRef]
7. Aliofkhazraei, M.; Ali, N.; Milne, W.I.; Ozkan, C.S.; Mitura, S.; Gervasoni, J.L. *Graphene Science Handbook: Fabrication Methods*; CRC Press: Boca Raton, FL, USA, 2016; pp. 3–179. ISBN 9781466591288.
8. Hummers, W.S.; Offeman, R.E. Preparation of Graphitic Oxide. *J. Am. Chem. Soc.* **1958**, *80*, 1339. [CrossRef]
9. Titelman, G.I.; Gelman, V.; Bron, S.; Khalfin, R.L.; Cohen, Y.; Bianco-Peled, H. Characteristics and microstructure of aqueous colloidal dispersions of graphite oxide. *Carbon* **2005**, *43*, 641–649. [CrossRef]
10. Paredes, J.I.; Villar-Rodil, S.; Solís-Fernandez, P.; Fernandez-Merino, M.J.; Guardia, L.; Martínez-Alonso, A.; Tascon, J.M.D. Preparation, characterization and fundamental studies on graphenes by liquid-phase processing of graphite. *J. Alloys Compd.* **2012**, *536*, S450–S455. [CrossRef]
11. Khan, U.; Porwal, H.; O'Neill, A.; Nawaz, K.; May, P.; Coleman, J.N. Solvent-exfoliated graphene at extremely high concentration. *Langmuir* **2011**, *27*, 9077–9082. [CrossRef] [PubMed]
12. Lerf, A.; He, H.; Forster, M.; Klinowski, J. Structure of graphite oxide revisited. *J. Phys. Chem. B* **1998**, *102*, 4477–4482. [CrossRef]
13. Park, J.; Yan, M. Covalent functionalization of graphene with reactive intermediates. *Acc. Chem. Res.* **2013**, *46*, 181–189. [CrossRef] [PubMed]
14. Dreyer, D.R.; Park, S.; Bielawski, C.W.; Ruoff, R.S. The chemistry of graphene oxide. *Chem. Soc. Rev.* **2010**, *39*, 228–240. [CrossRef]

15. Kuila, T.; Bose, S.; Mishra, A.K.; Khanra, P.; Kim, N.H.; Lee, J.H. Chemical functionalization of graphene and its applications. *Prog. Mater. Sci.* **2012**, *57*, 1061–1105. [CrossRef]

16. Jianchang, L.; Xiangqiong, Z.; Tianhui, R.; van der Heide, E. The preparation of graphene oxide and its derivatives and their application in bio-tribological systems. *Lubricants* **2014**, *2*, 137–161. [CrossRef]

17. Wang, G.; Wang, B.; Park, J.; Yang, J.; Shen, X.; Yao, J. Synthesis of enhanced hydrophilic and hydrophobic graphene oxide nanosheets by a solvothermal method. *Carbon* **2009**, *47*, 68–72. [CrossRef]

18. Veca, L.M.; Lu, F.; Meziani, M.J.; Cao, L.; Zhang, P.; Qi, G.; Qu, L.; Shrestha, M.; Sun, Y.-P. Polymer functionalization and solubilization of carbon nanosheets. *Chem. Commun.* **2009**, *18*, 2565–2567. [CrossRef] [PubMed]

19. Yang, Y.; Wang, J.; Zhang, J.; Liu, J.; Yang, X.; Zhao, H. Exfoliated graphite oxide decorated by PDMAEMA chains and polymer particles. *Langmuir* **2009**, *25*, 11808–11814. [CrossRef] [PubMed]

20. Stankovich, S.; Piner, R.D.; Nguyen, S.T.; Ruoff, R.S. Synthesis and exfoliation of isocyanate-treated graphene oxide nanoplatelets. *Carbon* **2006**, *44*, 3342–3347. [CrossRef]

21. Yang, H.; Shan, C.; Li, F.; Han, D.; Zhang, Q.; Niu, L. Covalent functionalization of polydisperse chemically-converted graphene sheets with amine-terminated ionic liquid. *Chem. Commun.* **2009**, *26*, 3880–3882. [CrossRef] [PubMed]

22. Díez-Pascual, A.M.; García-García, D.; San Andrés, M.P.; Vera, S. Determination of riboflavin based on fluorescence quenching by graphene dispersions in polyethylene glycol. *RSC Adv.* **2016**, *6*, 19686–19699. [CrossRef]

23. Smith, M.B.; March, J. *March's Advanced Organic Chemistry: Reactions, Mechanisms, and Structure*, 6th ed.; John Wiley & Sons Inc.: New York, NY, USA, 2001; pp. 1182–1183. ISBN 978-0-470-08494-6.

24. Sun, D.-L.; Mai, J.-J.; Deng, J.-R.; Idem, R.; Liang, Z.-W. One-Pot Synthesis of Dialkyl Hexane-1,6-Dicarbamate from 1,6-Hexanediamine, Urea, and Alcohol over Zinc-Incorporated Berlinite (ZnAlPO4). *Catalyst* **2016**, *6*, 28. [CrossRef]

25. Farkas, A.; Strohm, P.F. Mechanism of the Amine-Catalyzed Reaction of Isocyanates with Hydroxyl Compounds. *Ind. Eng. Chem. Fund.* **1965**, *4*, 32–38. [CrossRef]

26. Burkhart, G.; Kollmeier, H.J.; Schloens, H.H. The Importance of Catalysts for the Formation of Flexible Polyurethane Foams. *J. Cell. Plast.* **1984**, *20*, 37–41. [CrossRef]

27. Baker, J.W.; Holdsworth, J.B. The Mechanism of Aromatic Side-chain Reactions with Special Reference to the Polar Effects of Substituents. Part XIII. Kinetic Examination of the Reaction of Aryl Isocyanates with Methyl Alcohol. *J. Chem. Soc.* **1947**, 713–726. [CrossRef]

28. Mallakpour, S.; Yousefian, H. Reaction of aromatic carboxylic acids with isocyanates using ionic liquids as novel and efficient media. *J. Braz. Chem. Soc.* **2007**, *18*, 1220–1228. [CrossRef]

29. Uribe, M.; Hodd, K.A. The catalysed reaction of isocyanate and epoxide groups: A study using differential scanning calorimetry. *Thermochim. Acta* **1984**, *77*, 367–373. [CrossRef]

30. Dyer, E.; Taylor, H.A.; Mason, S.J.; Samson, J.; Amer, J. The Rates of Reaction of Isocyanates with Alcohols. I. Phenyl Isocyanate with 1- and 2-Butanol. *Chem. Soc.* **1949**, *71*, 4106–4109. [CrossRef]

31. Sato, M. The Rate of the Reaction of Isocyanates with Alcohols. II. *J. Org. Chem.* **1962**, *27*, 819–825. [CrossRef]

32. Burkus, J.; Eckert, C.F. The Kinetics of the Triethylamine-catalyzed reaction of diisocyanates with 1-butanol in toluene. *J. Am. Chem. Soc.* **1958**, *80*, 5948–5950. [CrossRef]

33. Ephrain, S.; Woodward, A.E.; Mesrobian, R.B. Kinetic Studies of the Reaction of Phenyl Isocyanate with Alcohols in Various Solvents. *J. Am. Chem. Soc.* **1958**, *80*, 1326–1328. [CrossRef]

34. Lu, Q.-W.; Hoye, T.R.; Makosko, C.W. Reactivity of common functional groups with urethanes: models for reactive compatibilization of thermoplastic polyurethane blends. *Polym. Chem.* **2002**, *40*, 2310–2328. [CrossRef]

35. Flipsen, T.A.C. Design, Synthesis and Properties of New Materials Based on Densely Crosslinked Polymers for Polymer Optical Fiber and Amplifier Applications. Ph.D. Thesis, University of Groningen, Groningen, The Netherlands, 2000.

36. Delebecq, E.; Pascault, J.-P.; Boutevinand, B.; Ganachaud, F. On the Versatility of Urethane/Urea Bonds: Reversibility, Blocked Isocyanate, and Non-isocyanate Polyurethane. *Chem. Rev.* **2013**, *113*, 80–118. [CrossRef] [PubMed]

37. Bastos, E.L.; da Silva, S.M.; Baader, W.J. Solvent Cage Effects: Basis of a General Mechanism for Efficient Chemiluminescence. *J. Org. Chem.* **2013**, *78*, 4432–4439. [CrossRef] [PubMed]

38. Lotya, M.; King, P.J.; Khan, U.; De, S.; Coleman, J.N. High-concentration, surfactant-stabilized graphene dispersions. *ACS Nano* **2010**, *4*, 3155–3162. [CrossRef] [PubMed]

39. Szabó, T.; Berkesi, O.; Dékány, I. DRIFT study of deuterium-exchanged graphite oxide. *Carbon* **2005**, *43*, 3186–3189. [CrossRef]

40. Lin-Ven, D.; Colthup, N.B.; Fateley, W.G.; Grasselli, J.G. *The Handbook of Infrared and Raman Characteristic Frequencies of Organic Molecules*; Academic Press: San Diego, CA, USA, 1991; pp. 155–225. ISBN 9780080571164.

41. Cannon, C.G. Infrared frequencies of amide, urea, and urethane groups. *J. Phys. Chem.* **1976**, *80*, 1247–1248. [CrossRef]

42. Díez-Pascual, A.M.; Díez-Vicente, A.L. Poly(3-hydroxybutyrate)/ZnO Bionanocomposites with Improved Mechanical, Barrier and Antibacterial Properties. *Int. J. Mol. Sci.* **2014**, *15*, 10950–10973. [CrossRef] [PubMed]

43. Mehrdad Mahkam, M.; Doostie, L. The Relation Between Swelling Properties and Cross-Linking of Hydrogels Designed for Colon Specific Drug Delivery. *Drug Deliv.* **2005**, *12*, 343–347. [CrossRef] [PubMed]

44. Xu, L.-C.; Siedlecki, C.A. Effects of surface wettability and contact time on protein adhesion to biomaterial surfaces. *Biomaterials* **2007**, *28*, 3273–3283. [CrossRef] [PubMed]

45. Konios, D.; Stylianakis, M.M.; Stratakis, E.; Kymakis, E. Dispersion behavior of graphene oxide and reduced graphene oxide. *J. Colloid Interface Sci.* **2014**, *430*, 108–112. [CrossRef] [PubMed]

46. Villar-Rodil, S.; Paredes, J.I.; Martínez-Alonso, A.; Tascón, J.M.D. Preparation of graphene dispersions and graphene-polymer composites in organic media. *J. Mater. Chem.* **2009**, *19*, 3591–3593. [CrossRef]

47. Wang, S.; Zhang, Y.; Abidi, N.; Cabrales, L. Wettability and Surface Free Energy of Graphene Films. *Langmuir* **2009**, *25*, 11078–11081. [CrossRef] [PubMed]

48. Huang, Y.; Yan, W.; Xu, Y.; Huang, L.; Chen, Y. Functionalization of graphene oxide by two step alkylation. In *Chemical Synthesis and Applications of Graphene and Carbon Materials*; Antonietti, M., Klaus Müllen, K., Eds.; Wiley-VCH Verlag: Weinheim, Germany, 2017; pp. 43–53. ISBN 978-3-527-33208-3.

49. Hu, H.; Allan, C.C.K.; Li, J.; Kong, Y.; Wang, X.; Xin, J.H.; Hu, H. Multifunctional organically modified graphene with super-hydrophobicity. *Nano Res.* **2014**, *7*, 418–433. [CrossRef]

50. Marques, P.; Gonçalves, G.; Cruz, S.; Almeida, N.; Singh, M.; Grácio, J.; Sousa, A. Functionalized Graphene Nanocomposites. In *Advances in Nanocomposite Technology*; Hashim, A.A., Ed.; IntechOpen: London, UK, 2011; pp. 247–272. ISBN 978-953-307-347-7.

51. Diez-Pascual, A.M.; Valles, C.; Mateos, R.; Vera-López, S.; Kinloch, I.A.; San Andrés, M.P. Influence of surfactants of different nature and chain length on the morphology, thermal stability and sheet resistance of graphene. *Soft Matter* **2018**, *14*, 6013–6023. [CrossRef] [PubMed]

52. Ossonon, B.D.; Bélanger, D. Synthesis and characterization of sulfophenyl-functionalized reduced graphene oxide sheets. *RSC Adv.* **2017**, *7*, 27224–27234. [CrossRef]

53. Namvari, M.; Biswas, C.S.; Galluzzi, M.; Wang, Q.; Du, B.; Stadler, F.J. Reduced graphene oxide composites with water soluble copolymers having tailored lower critical solution temperatures and unique tube-like structure. *Sci. Rep.* **2017**, *7*, 44508. [CrossRef] [PubMed]

54. Layek, R.K.; Nandi, A.K. A review on synthesis and properties of polymer functionalized graphene. *Polymer* **2013**, *54*, 5087–5103. [CrossRef]

55. Ferrari, A.C.; Robertson, J. Interpretation of Raman spectra of disordered and amorphous carbon. *Phys. Rev. B* **2000**, *61*, 14095–14107. [CrossRef]

56. Zhang, D.; Yang, S.; Chen, Y.; Liu, S.; Zhao, H.; Gu, J. ^{60}Co γ-ray Irradiation Crosslinking of Chitosan/Graphene Oxide Composite Film: Swelling, Thermal Stability, Mechanical, and Antibacterial Properties. *Polymers* **2018**, *10*, 294. [CrossRef]

MDPI

St. Alban-Anlage 66

4052 Basel

Switzerland

Tel. +41 61 683 77 34

Fax +41 61 302 89 18

www.mdpi.com

Nanomaterials Editorial Office

E-mail: nanomaterials@mdpi.com

www.mdpi.com/journal/nanomaterials